Lecture Notes in Mathematics

Edited by A. Dold and B. Eckmann

558

Boris Weisfeiler

On Construction and Identification of Graphs

Springer-Verlag
Berlin · Heidelberg · New York 1976

Boris Weisfeiler
Department of Mathematics
Pennsylvania State University
University Park, PA 16802

Library of Congress Cataloging in Publication Data
Main entry under title:

On construction and identification of graphs.

(Lecture notes in mathematics ; 558)
Bibliography: p.
Includes indexes.
1. Graph theory--Addresses, essays, lectures.
I. Weisfeiler, Boris, 1941- II. Series: Lecture notes in mathematics (Berlin) ; 558.
QA3.L28 no. 558 [QA166] 510'.8s [511'.5] 76-54232

AMS Subject Classifications (1970): 05CXX, 05B30, 05-04

ISBN 3-540-08051-1 Springer-Verlag Berlin · Heidelberg · New York
ISBN 0-387-08051-1 Springer-Verlag New York · Heidelberg · Berlin

Printed in Germany

Printing and binding: Beltz Offsetdruck, Hemsbach/Bergstr.

В целях природы обуздания
В целях рассеять неученья
Тьму
Берём картину мироздания - да!
И тупо смотрим, что к чему...
(А. и Б. Стругацкие)

... en vérité le chemin importe
peu, la volonté d'arriver suffit
a tout.
(A. Camus)

Acknowledgement

I am greatly indebted to Professor J. J. Seidel. His interest in the problems discussed in this volume encouraged me and his criticism, both mathematical and linguistical, helped to make the manuscript better. I am also grateful to Professor D. G. Corneil, who sent me many preprints of his work, and to J. Hager for linguistical corrections. It is a pleasure to thank here C. D. Underwood, the Administrative Officer of the School of Mathematics, I.A.S., for her patience and help during my stay at the Institute, and the Secretaries, I. C. Abagnale, A. W. Becker, E. T. Laurent, and M. M. Murray for their excellent typing. I am also grateful to the Institute for Advanced Study for support during my stay there.

B. Weisfeiler

TABLE OF CONTENTS

INTERDEPENDENCE OF SECTIONS

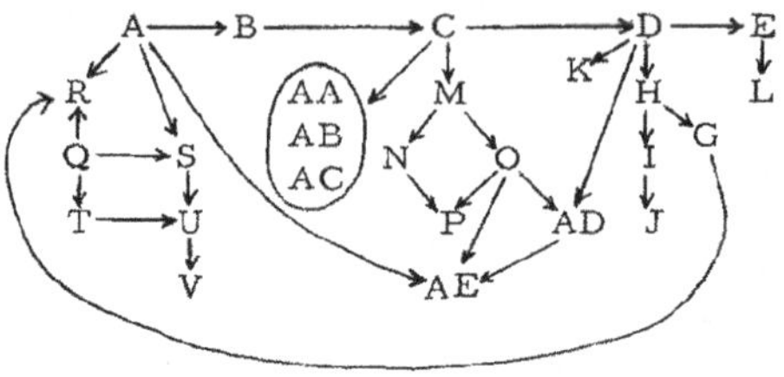

Section F provides motivations and examples for Sections D, E, G, H, I, J, K, L, N, O, P.

INTRODUCTION

In this volume we give an exposition of some results and introduce some notions which were encountered during attempts to find a good method of graph identification.

Sections of this volume are based mostly on unpublished papers of different people. I ask the reader who wishes to refer to papers constituting this volume to refer to them by the names given in the Table of Contents. Papers which are not followed by any name can be cited as my own.

The beginning of our work was the research described in [We 3]. It was shown in this paper how to put into correspondence with any graph a nice combinatorial object. The authors were not conscious at the time of the writing [We 3] that this combinatorial object was related to other problems. Later it turned out that the same object had been independently discovered and studied in detail by D. G. Higman [Hi 3], [Hi 5], [Hi 6] and that such formations as strongly regular graphs, symmetric block designs, centralizer rings of permutations groups are special cases of this object (cf., Section F and L18).

Although the properties of this object, called here a cellular algebra, were discussed by D. G. Higman [Hi 3], [Hi 6], we decided to state here some assertions about them. This is done in the hope that it will help a reader to get acquainted with the notions and their use.

At the same time the main stress is on the description of operations and constructions. Some assertions are proved to show how these constructions work.

In an attempt to acquire a new understanding of the nature of our problems, much practical work was done, mostly with the help of computers. The most interesting outcome in this direction is probably the program which was designed to generate all strongly regular graphs with $\leq$ 32 vertices. This program constructed all strongly regular graphs with 25, 26, 28 vertices, but failed, for lack of time, to construct such graphs with 29 vertices. This work is described in Sections S-V. The strongly regular graphs with 25 and 26 vertices were intensively studied (cf., e.g., [Se 5], [Sh 4]).

Let us give now a brief description of the content of this volume.

We begin with a discussion of certain questions connected with the graph isomorphism problem (Section A). Then we show in Section B how the development of known and natural approaches leads to our main construction which is described in detail in Section C. This construction gives rise to the notion of cellular algebras. We discuss properties of cellular algebras in Sections D and E.

We show then that centralizer rings of permutation group theory are cellular algebras (Section F) and describe in Section G some general classes of cellular algebras. The constructions of Section G are modeled on permutation group theory.

Sections H-K deal with imprimitivity and primitivity of cellular algebras. These classes of cellular algebras arise naturally when one tries to describe the structure of general cellular algebras and they are analogous to the corresponding notions of permutation group theory.

In Section L some arithmetical relations between the numerical parameters of cellular algebras are obtained with help of algebraic theory. This section shows that the algebra structure can be used to get combinatorial information. Recent D. G. Higman's results [Hi 5], [Hi 6] cover most results of this Section.

In Section M we pass to the more algorithmic point of view. But otherwise it is essentially a repetition of Section C. In Section N and O new operations on graphs are introduced, and it is shown how the stability with respect to these operations restricts the structure of a cellular algebra. In Section P we show that the stability of a cellular algebra under some set of operations can be used to prove results which hold for centralizer rings of permutation groups.

In Section Q we describe our setup and terminology before proceeding to the study of algorithms. These algorithms are described in Sections R-T. In Section U the results of the program based on the algorithms of Section T are presented and the information based on these results is discussed.

In the Appendix (Sections AA-AE; the first A stands for "Appendix") we discuss different applications of the notions introduced in the main part of this volume.

CONVENTIONS, ASSUMPTIONS, NOTATIONS.

1. The references in this book are organized in the following manner: Sections are numbered by capital Roman letters; references inside one section do not use indication of the section; references to other sections begin with the letter (or letters) of the section. If several references to one section are written successively, they are divided by commas and the name of the section is used usually once.

E.g., in Section L, references 4.1, 3.2; K15, 16; B7.6.15 mean that subsections 4.1, 3.2 of Section L, subsections 15, 16 of Section K, subsection 7.6.15 of Section B are referred to.

References to original papers begin with the first two letters of the name of the author.

2. Assumptions and Peculiarities of Terminology.

The word "graph" is used in two different senses: one is the usual notion of a graph; for the second one, see C1.

A simple graph is a graph without loops, multiple or directed edges. The valency of a vertex of such a graph is the number of edges incident to this vertex.

In Sections M-O, R a partial order satisfies an additional condition of Section M2.1.

By the composition of a matrix A we mean the list of different entries of A together with their multiplicities. E.g., if $A = xI_{m,n}$, then A is composed of x with the multiplicity mn. We say that the compositions of A and B are disjoint (or that A and B are disjoint) if A and B have no common entries. We say that A and B have the same or equal composition if A and B are composed of the same entries with the same multiplicities. If A and B are matrices whose entries belong to a partially ordered set M, we say that (the composition of A) is greater than (the composition of B), or simply $A \geq B$ if this holds for compositions. Here the members of the composition of a matrix are ordered correspondingly to the order in M and comparison is understood lexicographically.

If $A = (a_{ij})$ is an $(n \times m)$-matrix, $g \in \mathrm{Sym}\, n$, $h \in \mathrm{Sym}\, m$, then $g^{-1}Ah = (a_{gi,hj})$.

Capital German letters $\mathfrak{A}$, $\mathfrak{B}$, $\mathfrak{C}$ usually denote a cellular algebra or a normal subcell.

Letters X, Y, Z usually denote a graph (a matrix whose entries are independent variables). Letters U, V, W usually denote a set of points.

For typing reasons we write sometimes Σ_s or $\prod_{s \in T}$, etc., for $\sum\limits_{s}$ or $\prod\limits_{s \in T}$, etc.

3. General Notations.

E_n - identity matrix.

$I_{m,n}$ - $(m \times n)$-matrix all of whose entries are ones.

$I_n = I_{n,n}$.

$\tilde{I}_n = I_n - E_n$.

$i_n = I_{n,1}$.

$\mathrm{diag}(a_1, \ldots, a_n)$ - diagonal $(n \times n)$-matrix with $a_1, \ldots, a_n$ as diagonal entries.

$h_i = \mathrm{diag}(0, \ldots, 0, 1, 0, \ldots, 0)$, with 1 at the position i.

$$A \otimes B = \begin{pmatrix} a_{11}B & a_{12}B & \ldots \\ a_{21}B & \ldots & \\ \ldots & & \end{pmatrix},$$ where A, B are (possibly) rectangular matrices and $A = (a_{ij})$.

$\mathrm{Sp}\,A$ - trace of A.

$\dim A$ - number of different entries of $(m\times n)$-matrix A, e.g., $\dim\begin{pmatrix}1&2\\1&1\end{pmatrix} = 2$, $\dim I_{m,n} = 1$.

$|A|$ - degree (i.e., n) of a square $(n\times n)$-matrix A.

A' - transposed of A.

$S_n = xE_n + y\tilde{I}_n$ - simplex.

$A = R$ (said: A is split) means that $\dim A = n^2$, with $|A| = n$.

$A = \mathrm{const}$ (said: A is constant) denotes that $\dim A = 1$, i.e., $A = xI_{m,n}$ for appropriate x, m, n.

$d(C)$ - the number of ones in any non-zero row of a $(0,1)$-matrix (is applied only when it does not depend on the row).

$A \subseteq B$ means that $b_{ij} = b_{kd}$ implies $a_{ij} = a_{kd}$ where $A = (a_{ij})$, $B = (b_{ij})$.

If X is an $(m\times n)$-matrix, M a submatrix (i.e., a subset of the set of mn positions (i,j)), and e, f are $(0,1)$-matrices, then

$e \subset M$ denotes that all ones of e lie in M;

$e \cap M \neq 0$ denotes that some ones of e lie in M;

$e \cap M = 0$ denotes that M contains only zeros of e;

$e \subset f$ (resp. $e \cap f \neq 0$, resp. $e \cap f = 0$) denotes that all (resp. some, resp. none) ones of e are ones of f.

$A(V,W)$ is the submatrix of A cut out by rows with numbers in V and columns with numbers in W, that is, if $A = (a_{ij})$, then $A(V,W) = (a_{ij})_{i\in V,\, j\in W}$.

$\mathbb{Z}$, $\mathbb{N}$, $\mathbb{Q}$, $\mathbb{R}$, $\mathbb{C}$ denote the set of the integers, positive integers, rational, real, complex numbers.

$\mathbb{Z}^+ = \mathbb{N} \cup 0$.

$[m, n] = \{m, m+1, \ldots, n\}$.

(m, n) = greatest common divisor of m and n.

$|V|$ - cardinality of a set V.

Sym n, Sym(n) - symmetric group of all permutations of n symbols.

Sym V, Sym(V) - group of all permutations of a set V.

A. SOME REMARKS ABOUT THE PROBLEM OF GRAPH IDENTIFICATION.

An algorithm of graph identification is an algorithm $\mathcal{A}$ whose domain consists of pairs of graphs and whose result on a pair Γ_1, Γ_2 is +1 if Γ_1 is isomorphic to Γ_2 and -1 if not. Let us associate with $\mathcal{A}$ the function ("speed") $f(\mathcal{A}, n)$, whose value at n is the maximum number of steps required by $\mathcal{A}$, in order to find the result for any pair of graphs Γ_1, Γ_2 with n vertices.

The problem of graph identification is to find an algorithm of graph identification which for any other such algorithm $\mathcal{B}$ yields

$$f(\mathcal{A}, n) \leq f(\mathcal{B}, n) ,$$

for all sufficiently large n.

The evident algorithm requires n! steps. It is not clear whether the function $f(\mathcal{A}, n)$ for an "optimal" algorithm $\mathcal{A}$ is polynomial or not. Possibly, for any constant b there is no algorithm $\mathcal{B}$ such that $f(\mathcal{B}, n) \leq n^b$ for all $n >> 0$. But in any case, as far as I know, there is even no algorithm for which it is proved that

$$f(\mathcal{A}, n) \cdot 2^{-cn} \to 0 \text{ for } c \geq 0.$$

For special classes of graphs such as trees or planar graphs, the situation is much better. In these cases there exist algorithms which achieve theoretical lower bounds on the number $f(\mathcal{A}, n)$ ([Ho 3], [Sk 1]).

In the general case there are at present some results which show that the situation in some close problems is almost hopeless (cf. [Ka 1]).

There are two directions in the history of published approaches to the problem of graph identification; let us call them conditionally "local" and "global".

In the global approach (e. g. , [Va 1], [Li 1], [Li 3], [Tu 1]) the tried different algebraic invariants of the adjacency matrices of the given graphs. Most common here are the characteristic polynomial, the permanent, et al. There is also a lot of literature where these invariants are shown to be insufficient. However, there are many invariants (cf. Section AE), and a responsible approach should consist of proving that the given invariants distinguish the graphs and could themselves be computed sufficiently fast. Related questions are discussed in more detail in Section AE.

In the local approach one tries to construct a sufficient number of invariants of every vertex of the given graphs in terms of configurations containing fixed (say, 2 or 3) numbers of points and passing through the given vertex. This approach is the oldest one (cf. [Na 1], [Un 1], [Mo 1]—all ten years old). These authors used configurations with ≤ 2 points (i. e. , they used edges). However, a recursive application of this approach (cf. [Ba 1], [Sk 1]) can lead to more information than at first glance would seem possible (for more details cf. Sections B, C).

The next step is to consider configurations with 3 points. Here we also have many papers (e. g., [We 3], [Le 2]). A. A. Lehmann and B. Weisfeiler's joint paper [We 3] (cf. also Section R) appears to present the best algorithm. Then we thought that it was time to stop and think. Indeed, the object which was constructed with the help of configurations of size 3 is very nice, which probably implies that it is natural and that our approach up to this moment was a correct one. However, nothing nice is seen before us or around us which implies (also probably) that we have to search further for a right road.

These geometrical approaches are discussed in more detail in the next section. Some examples are also given there. The aim of all of them is to construct a partition of the vertices of the given graph into orbits under the automorphism group of this graph.

One more merit of these approaches is that one is forced to study graphs, and even if a good algorithm is not found, one can still hope to find interesting objects or unconventional results.

Anyway, now we still have to make an exhaustive search. Whatever refinements and improvements we have made only make this search "somewhat" shorter, but have not replaced it. In an exhaustive search we fix, in turn, all vertices of the group of vertices having the same number of configurations of certain given types. Then to the resulting graphs with one fixed vertex we again apply our local geometrical approach. And so forth. There is no reason to avoid doing this. However, if we do it too many times (of order n, say) then this would mean that our algorithm requires 2^{cn} steps and in a sense is as good as the usual exhaustion. So the question is: What is the depth of our exhaustive search? This question has not yet been non-trivially answered in any version of an algorithm of graph identification.

Possibly in the absence of a good algorithm one can prove that this algorithm is statistically good in some sense. For example, it would be nice (in any case, with or without an estimate) to know the function $F(\mathcal{A}, n, b)$, that is, the number of pairs of graphs with n vertices for which $\mathcal{A}$ computes the result in $\leq n^b$ steps.

In the geometrical approach one tries, de facto, to find a canonical numeration of the vertices of the given graph and then to compare the results for two of them. This procedure is usually disguised by making comparison after each step of canonization. The algorithms we describe in Sections R and S are algorithms of graph canonization.

This approach is better than the usual graph identification if one has many graphs to compare (as, for example, the algorithm of Section S which worked on results of algorithm of Section T). Namely, one has to canonize every graph and to keep only different canonical forms. So in place of $\binom{n}{2}$ applications of an algorithm of graph identification, one can use n times an algorithm of graph canonization and then make $\binom{n}{2}$ (or less) comparisons. Of course, this approach is an unworthy one if one has a good algorithm of identification and a bad algorithm of canonization.

B. MOTIVATION.

We discuss below steps which lead naturally to our main formalism. The resulting construction permits us to associate with any finite graph Γ a matrix algebra which is uniquely determined by the graph up to permutation of the elements of the basis. This construction generalizes and develops different algorithms used to approach the graph isomorphism problem. Here are some examples of such algorithms.

1. Summation of the Weights of Vertices over Neighbours (e.g., [Mo 1]). Suppose we are given a simple graph Γ. The procedure is iterative. In the first step every vertex is given weight 1 and all vertices form one unique class. Suppose that in some of the later steps we have some partition $V = V(\Gamma) = \bigcup V_i$ and the vertices of each V_i have the same weights. In the next step we take the sum of the weights of all vertices adjacent to the given one as the new weight of the given vertex. The subsets of the new partition of V are the sets of all vertices where the function of weight is a constant. The process stops if we obtain no new partitions.

Example:

Γ : 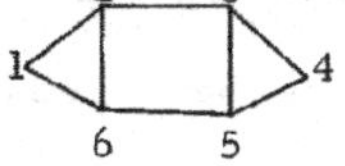

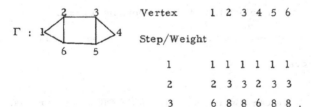

Vertex	1	2	3	4	5	6
Step/Weight						
1	1	1	1	1	1	1
2	2	3	3	2	3	3
3	6	8	8	6	8	8 .

Therefore, the stabilization occurs at the second step and the partition of vertices is (1,4), (2,3,6,5).

2. Summation of the Weights over a Partition of the Vertices (e.g., [Sk 1]). In this case one associates with a vertex a vector of weights. The number

of coordinates of this vector is the number of subsets into which $V = V(\Gamma)$ is partitioned.

In the first step (as in 1 above) every vertex is given weight 1, and the partition is trivial (it consists only of V). Suppose now that we have some partition $V = \bigcup V_i$. Then the weight of a vertex $v \in V$ in the next step is the vector whose i-th component is the sum of the valencies of all vertices which belong to the i-th class V_i and which are adjacent to v. The subsets of the new partition are those subsets where the weight is constant. These subsets are numbered according to the (dictionary) order of weights they represent. The process stops when there are no new partitions.

Examples:

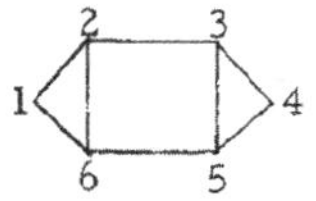

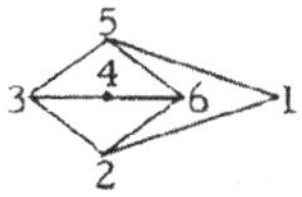

Vertex	1	2	3	4	5	6	1	2	3	4	5	6
Step/Weight												
1	1	1	1	1	1	1	1	1	1	1	1	1
2	2	3	3	2	3	3	2	3	3	2	3	3
3	(0,2)	(1,2)	(1,2)	(0,2)	(1,2)	(1,2)	(0,2)	(1,2)	(1,2)	(0,2)	(1,2)	(1,2) .

It is not possible to do more with these graphs since the achieved partition is a partition into orbits of the automorphism group.

Note however that for regular graphs these methods will not give a partition of vertices.

3. We can now try to partition the edges of the graphs. As a first approximation, we can consider the number of vertices incident to both vertices of the edge. In the first example above we have two edges which are contained in triangles. In the second example there are no such edges. Therefore, this pair of graphs is not isomorphic, although there is no distinction of the vector weights of

the vertices. However, the following graph

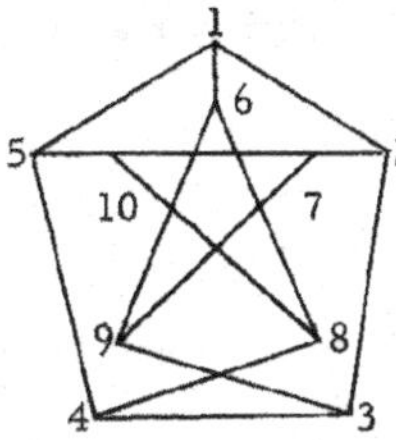

is also immune to this procedure. Nevertheless, it can be seen that the automorphism group of this graph is not transitive.

4. To further strengthen the procedure for detecting the differences of vertices and edges, we can consider not only the edges of the given graph but also the edges of its complement $\bar{\Gamma}$. (Recall that vertices of $\bar{\Gamma}$ are vertices of Γ, and edges of $\bar{\Gamma}$ are non-edges of Γ.) In the above example of the graph Γ (with 10 vertices) the use of $\bar{\Gamma}$ permits us to distinguish the pair 1, 6 of vertices. Namely, any edge of the graph $\bar{\Gamma}$ incident to these vertices is contained in <u>one</u> triangle with two sides in Γ and one (given) side in $\bar{\Gamma}$. On the other hand, the edges of $\bar{\Gamma}$ incident to remaining vertices are partitioned into three classes according to the number of triangles which contain a given edge and whose two sides are in Γ. For instance, for vertex 2 these classes are:

edge (2, 8) is not contained in a triangle with two sides in Γ;

edges (2, 4), (2, 5), (2, 6), (2, 10) are contained in one triangle each;

edge (2, 9) is contained in two triangles.

5. Once we began to distinguish edges and vertices, we have to use convenient and effective machinery to describe this. We use the following formalization. Instead of the adjacency matrix of a simple graph Γ we consider the matrix $X = X(\Gamma)$ whose elements are independent variables. We replace

the ones by one variable x, say; the non-diagonal zeros by another variable, say y; and the diagonal zeros by a third variable, say z.

Now the process described in 1 above consists simply of taking the sum of the entries of X over its rows.

The process described in 2 can be described as a reconstruction of X. Namely, first the diagonal elements of X are changed according to the respective row-sums. Call the new matrix Y then we have, for $X = (x_{ij})$, $Y = (y_{ij})$,

$$y_{ii} = y_{jj} \Leftrightarrow \sum_k x_{ik} = \sum_k x_{jk}.$$

This means that we get a partition of vertices. Now edges joining vertices of different classes also belong to different classes. So we require, next, that

$$y_{ij} = y_{k\ell} \Rightarrow y_{ii} = y_{kk}, y_{jj} = y_{\ell\ell}.$$

Stabilization in 2 above corresponds to iteration of this construction.

5.1. Remark. The above description is not an algorithm because for an algorithm one needs to introduce some ordering (cf. Section M). We hope that the present discussion is sufficient for introductory purposes.

5.2. Remark. Another virtue of this approach is that it works equally well for graphs with loops, multiple edges, etc. (cf. also C1, M2).

6. The same formalism is convenient for a description of step 4, designed to distinguish edges. Note that if we consider the square of a matrix X, then the (i, j)-entry of this square describes the set of paths of length 2 from the vertex i to j. If, moreover, we assume that the variables of X do not

commute, this (i, j)-th entry describes the set of ordered paths. So the next step is the following. We consider $X^2 = (z_{ij})$, and we construct the matrix $Y = (y_{ij})$ using the rule

$$y_{ij} = y_{k\ell} \iff z_{ij} = z_{k\ell} .$$

(Here again one has to use some ordering, cf. Section M, but we disregard this for a moment.)

7. Let us show how this approach works for the graph Γ considered in 4. In the matrices X, Y, Z we shall write the indices of independent variables in place of independent variables.

X

1	2	3	3	2	2	3	3	3	3
2	1	2	3	3	3	2	3	3	3
3	2	1	2	3	3	3	3	2	3
3	3	2	1	2	3	3	2	3	3
2	3	3	2	1	3	3	3	3	2
2	3	3	3	3	1	3	2	2	3
3	2	3	3	3	3	1	3	2	2
3	3	3	2	3	2	3	1	3	2
3	3	2	3	3	2	2	3	1	3
3	3	3	3	2	3	2	2	3	1

$Y \leftrightarrow X^2$

1	2	3	3	2	2	3	3	3	3
2	1	2	3	3	3	2	4	5	3
3	2	1	2	3	3	5	3	2	4
3	3	2	1	2	3	4	2	3	5
2	3	3	2	1	3	3	5	4	2
2	3	3	3	3	1	3	2	2	3
3	2	5	4	3	3	1	3	2	2
3	4	3	2	5	2	3	1	3	2
3	5	2	3	4	2	2	3	1	3
3	3	4	5	2	3	2	2	3	1

$Z \leftrightarrow Y^2$

1	3	4	4	3	5	4	6	6	4
7	2	10	11	12	9	10	14	15	11
8	10	2	13	11	8	16	11	10	17
8	11	13	2	10	8	17	10	11	16
7	12	11	10	2	9	11	15	14	10
5	6	4	4	6	1	4	3	3	4
8	10	16	17	11	8	2	11	10	13
9	14	11	10	15	7	11	2	12	10
9	15	10	11	14	7	10	12	2	11
8	11	17	16	10	8	13	10	11	2

Y:

$x_1 : x_1x_1 + 3x_2x_2 + 6x_3x_3$

$x_2 : x_1x_2 + x_2x_1 + 2x_3x_2 + 2x_2x_3 + 4x_3x_3$

$x_3 : x_1x_3 + x_3x_1 + x_2x_2 + 2x_3x_2 + 2x_2x_3 + 3x_3x_3$

$x_4 : x_1x_3 + x_3x_1 + 3x_2x_3 + 3x_3x_2 + 2x_3x_3$

$x_5 : x_1x_3 + x_3x_1 + 2x_2x_2 + x_2x_3 + x_3x_2 + 4x_3x_3$

Z:

$x_1 : x_1x_1 + 3x_2x_2 + 6x_3x_3$

$x_2 : x_1x_1 + 3x_2x_2 + 4x_2x_3 + x_4x_4 + x_5x_5$

$x_3 : x_1x_2 + x_2x_1 + 2x_3x_2 + 2x_2x_3 + 2x_3x_3 + x_3x_4 + x_3x_5$

$x_4 : x_1x_3 + x_2x_2 + x_3x_1 + 2x_3x_2 + 2x_2x_3 + x_3x_3 + x_3x_4 + x_3x_5$

$x_5 : x_1x_2 + x_2x_1 + 2x_2x_3 + 2x_3x_2 + 4x_3x_3$

$x_6 : x_1x_3 + x_3x_1 + x_2x_4 + x_2x_5 + 3x_3x_3 + 2x_3x_2 + x_2x_2$

$x_7 : x_1x_2 + x_2x_1 + 2x_2x_3 + 2x_3x_2 + 2x_3x_3 + x_4x_3 + x_5x_3$

$x_8 : x_1x_3 + x_3x_1 + x_2x_2 + 2x_2x_3 + 2x_3x_2 + x_3x_3 + x_4x_3 + x_5x_3$

$x_9 : x_1x_3 + x_3x_1 + x_4x_2 + x_5x_2 + 3x_3x_3 + 2x_2x_3 + x_2x_2$

$x_{10} : x_1x_2 + x_2x_1 + x_2x_3 + x_3x_2 + 2x_3x_3 + x_2x_5 + x_5x_2 + x_3x_4 + x_4x_3$

$x_{11} : x_1x_3 + x_3x_1 + x_2x_3 + x_3x_2 + x_2x_2 + x_3x_3 + x_2x_4 + x_4x_2 + x_2x_5 + x_5x_3$

$x_{12} : x_1x_3 + x_3x_1 + x_2x_2 + 2x_2x_3 + 2x_3x_2 + x_3x_3 + x_4x_5 + x_5x_4$

$x_{13} : x_1x_2 + x_2x_1 + 2x_3x_3 + 2x_2x_3 + 2x_3x_2 + x_5x_4 + x_4x_5$

$x_{14} : x_1x_4 + x_4x_1 + 2x_2x_3 + 3x_3x_2 + x_3x_5 + x_5x_3$

$x_{15} : x_1x_5 + x_5x_1 + 2x_2x_2 + 2x_3x_3 + x_2x_3 + x_3x_2 + x_3x_4 + x_4x_3$

$x_{16} : x_1x_5 + x_5x_1 + 2x_2x_2 + 4x_3x_3 + x_2x_4 + x_4x_2$

$x_{17} : x_1x_4 + x_4x_1 + 2x_3x_3 + 2x_2x_3 + 2x_3x_2 + x_2x_5 + x_5x_2$

The matrix X contains three variables; x_1 is for the diagonal entries, x_2 is for the edges of the graph Γ, x_3 is for edges of the complementary graph of Γ. The variables of Y correspond to five different polynomials which are the entries of the matrix X^2. The square Y^2 of Y already contains 17 different polynomials; to each of them there corresponds an independent variable of the matrix Z. If, finally, one considers Z^2, one sees that diagonal variables are partitioned into three classes (1, 6), (2, 5, 8, 9) and (3, 4, 7, 10) and further squaring does not lead to new partitions. The permutations (written cyclically) (2, 5)(3, 4)(7, 10)(8, 9) and (3, 7)(4, 10) and (1, 6)(2, 9)(5, 8) are automorphisms of the graph Γ. They generate a group which acts transitively on the vertices of each class (and also on the edges of each class). Thus we have revealed all differences of vertices and edges of the graph Γ.

Let us note that the application of the described procedure to a simple graph can lead to an "orientation" of certain edges. For instance, in the graph below

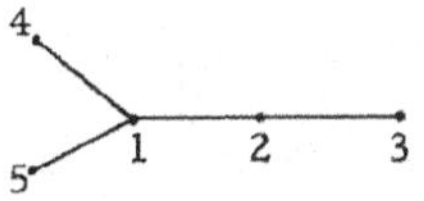

the edge (1, 2) can be considered as oriented (in the sense that its vertices are situated differently with respect to the whole graph). In Sections AA, AB, AC examples are given of simple graphs whose edges acquire "orientation" although the ends of the edges have no differences.

The use of matrix X, with independent variables as entries, permits one to employ the described procedure not only for simple graphs but also for oriented graphs, for graphs with multiple or coloured edges, etc. Thus, any graph is interpreted as a complete graph with some coloring of the edges and the vertices. This approach leads to a generalization of the definition of a graph (cf. C1).

Let us also note here that our definition generalizes the definition of A. A. Zykov [Zy 1] in the sense that in place of boolean rings we consider arbitrary rings. On the other hand, our definition is needed only to facilitate and formalize the exposition. All considerations might be (and sometimes are) also conducted in geometrical terms.

C. A CONSTRUCTION OF A STATIONARY GRAPH.

In this section we systematically describe the procedures introduced in the preceding section. The result of these procedures is an invariant of the given graph. This invariant is constructed in the same manner for all graphs (simple, with coloured, directed, or multiple edges, and with coloured vertices). But even if one begins with a simple graph, it can acquire orientation of edges, colouration of vertices, etc. (cf. the preceding section and Sections AA, AB, AC; in these sections one can also find examples of the application of the constructions of this section).

Since we are forced to consider quite different kinds of graphs, it is convenient to make the following definition.

1. Definition. An $(n \times n)$-matrix $X = (x_{ij})$ is called a graph if its entries are independent variables z_k, $k = 1, \dots, N$, and if $x_{ii} \neq x_{st}$ for $s \neq t$. The number n is called the degree of X and the number N of different variables which are entries of X is called the dimension of X. Notations: $n = |X|$, $N = \dim X$. We assume throughout that independent variables do not commute.

If a geometrical image of a graph is preferred, one can consider the complete graph with coloured vertices and edges. It can be assumed that at each vertex there is a loop, having the same colour as the vertex; this colour is coded in our matrix X by the corresponding diagonal entry. For each pair of vertices i and j there is either one undirected edge of colour x_{ij} (if $x_{ij} = x_{ji}$), or, if $x_{ij} \neq x_{ji}$, there is the directed edge of colour x_{ij} from i to j and the directed edge of colour x_{ji} from j to i.

2. Definitions. Let $X = (x_{ij})$ and $Y = (y_{ij})$ be two graphs of degree n. We say that:

2.1. A permutation matrix g of degree n is an isomorphism of X and Y if $gXg^{-1} = Y$. If $X = Y$, then g is called an automorphism of X. The set of all automorphisms of X is denoted Aut X.

2.2. X is imbedded in Y (denoted $X \subseteq Y$) if $y_{ij} = y_{k\ell}$ for all i, j, k, ℓ.

2.3. X is equivalent to Y (denoted $X \sim Y$) if $X \subseteq Y$ and $Y \subseteq X$.

2.4. An embedding $X \subseteq Y$ is canonical if $\mathrm{Aut}\,X = \mathrm{Aut}\,Y$.

To get the geometrical meaning of these definitions, suppose that equally numbered vertices of X and Y are identified. Then σ is an automorphism of X means that the pair of vertices before and after permutation σ are connected by an edge of the same colour. Further, X is imbedded in Y if equally coloured vertices of Y are also equally coloured in X.

3. Remarks.

3.1. If $X \subseteq Y$, then $\dim X \leq \dim Y$.

3.2. If $X \subseteq Y$, then $\mathrm{Aut}\,Y \subseteq \mathrm{Aut}\,X$.

3.3. If $X \subseteq Y$, $Y \subseteq Z$, then $X \subseteq Z$.

3.4. If $X \sim Y$, $Y \sim Z$, then $X \sim Z$. So $\sim$ is an equivalence relation.

3.5. Any graph of degree n contains the simplex S_n. This is the only graph of dimension 2; its matrix is $xE_n + y\tilde{I}_n$. In our approach we simultaneously consider several ordinary graphs; in the case of S_n these graphs are the complete graph and the empty graph. So "simplex" is the name for the equivalence class of these two graphs.

3.6. Every graph of degree n is imbedded in a graph of degree n and dimension n^2, which is unique up to equivalence. We call this graph the split graph and denote it by R.

4. Definitions. Let $X = (x_{ij})$ and $Y = (y_{ij})$ be two graphs of degree n. We say that:

4.1. The graph $Z = (z_{ij})$ is the superimposition of graphs X and Y (notation $Z = X \vee Y$) if

$$z_{ij} = z_{k\ell} \Longleftrightarrow x_{ij} = x_{k\ell} \text{ and } y_{ij} = y_{k\ell}.$$

4.2. The graph $Z = (z_{ij})$ is the product of X and Y (notation $Z = X \circ Y$) if

$$z_{ij} = z_{k\ell} \Longleftrightarrow \sum_s x_{is} y_{sj} = \sum_s x_{ks} y_{s\ell}.$$

(Recall that our variables do not commute.)

4.3. $\alpha(X) = (X \circ X) \vee (X \circ X)'$ is the extension of X (A' is the transpose of A).

4.4. X is stationary if $\alpha(X) \sim X$.

These definitions depend only on the equivalence class of X and Y, and the resulting graph of 4.1, 4.2, 4.3 is also defined up to equivalence.

To geometrically understand the meaning of the superimposition, one should imagine the colouring of an edge of $X \vee Y$ as the (ordered) mixture of the colourings of edges of X and Y.

In the case of the product, the colour of the edge between vertices i and j depends on the number and colouration of the paths of length 2 between vertices i and j such that the first edge of each path is an edge of X and the second one is an edge of Y. The polynomial $\sum_s x_{is} y_{sj}$ completely describes the set of these paths.

5. Lemma. Let $X = (x_{ij})$, $Y = (y_{ij})$, $Z = (z_{ij})$ be graphs of degree n and $Z = X \circ Y$. Then $x_{ii} \neq x_{jj}$ implies that $z_{ik} \neq z_{j\ell}$. Analogously $y_{ii} \neq y_{jj}$ implies that $z_{ki} \neq z_{\ell j}$ for all k, ℓ.

Proof. The second assertion is proved in the same way as the first one. So let us only prove the first one. We have to compare the set of paths of length 2 from the i-th to the k-th vertex with the analogous set from the j-th to the ℓ-th. Each set contains only one path beginning with a loop, namely, the first begins with the loop of the colour x_{ii} and then goes through the edge y_{ik}. The second one begins with the loop of colour x_{jj}. Since $x_{ii} \neq x_{jj}$, these sets of paths do not coincide, whence the assertion.

Formally, z_{ik} corresponds to $\sum_s x_{is}y_{sk}$ and $z_{j\ell}$ corresponds to $\sum_s x_{js}y_{s\ell}$. The first sum contains only summands $x_{ii}y_{ik}$ and $x_{ik}y_{kk}$ involving diagonal variables and the second only $x_{jj}y_{j\ell}$ and $x_{j\ell}y_{\ell\ell}$. Since by assumption $x_{ii} \neq x_{jj}$ and variables do not commute and since diagonal variables are different from non-diagonal ones by the definition of a graph, our assertion follows once more.

6. Lemma. Let X and Y be graphs of degree n.

6.1. $X \subseteq X \vee Y$, $Y \subseteq X \vee Y$.

6.2. $X \subseteq X \circ Y$, $Y \subseteq X \circ Y$.

Proof. 6.1 is evident. The second part of 6.2 is proved similarly to its first part. So let us prove the first part. We have to prove that $x_{ij} \neq x_{k\ell}$ implies $\sum x_{is}y_{sj} \neq \sum x_{ks}y_{s\ell}$. The only entries in these sums which contain the diagonal variables on the right are $x_{ij}y_{jj}$ and $x_{k\ell}y_{\ell\ell}$ respectively. Since $x_{ij} \neq x_{k\ell}$, the sums are different, which is our assertion.

7. Corollaries. Let X and Y be graphs of degree n.

7.1. $\mathrm{Aut}(X \vee Y) = \mathrm{Aut}\, X \cap \mathrm{Aut}\, Y$.

7.2. $\mathrm{Aut}(X \circ Y) = \mathrm{Aut}\, X \cap \mathrm{Aut}\, Y$.

7.3. The imbedding $X \subseteq X \circ X$ is a canonical one. In particular, X is canonically imbedded in $\alpha(X)$, $\mathrm{Aut}\, X = \mathrm{Aut}(X \circ X)$.

Proof. 7.1 and 7.2 follow from 6.1, 6.2 and 3.2; 7.3 follows from 7.1, 7.2.

8. Stabilization.

Let X be a graph of degree n. Put $X^{(0)} = X$, $X^{(i+1)} = \alpha(X^{(i)})$. By Lemma 6 we have $\dim X^{(i)} \leq \dim X^{(i+1)}$. On the other hand, we have $\dim X^{(i)} \leq n^2$ for all i. Since by 6.1, 6.2, $X^{(i)} \subseteq X^{(i+1)}$, we have $X^{(i)} \sim X^{(q)}$ for some q and for all $i \geq q$. Let us denote this graph $X^{(q)}$ by $\hat{X}$.

8.1. Lemma. Suppose X is a stationary graph and Y any graph. If $Y \subseteq X$, then $\hat{Y} \subseteq X$. In particular, $\mathrm{Aut}\, X \subseteq \mathrm{Aut}\, \hat{Y} = \mathrm{Aut}\, Y$.

This follows directly from Lemma 6 and Corollary 7.3.

From the above it follows:

8.2. Theorem. For any graph X there exists a unique (up to equivalence) stationary graph $\hat{X}$ such that X is canonically imbedded in $\hat{X}$. For every stationary graph Y such that X is canonically imbedded in Y one has $\hat{X} \subseteq Y$. In particular, for $\sigma \in \mathrm{Sym}(n)$, one has $\widehat{(\sigma X \sigma^{-1})} \sim \sigma \hat{X} \sigma^{-1}$.

9. Elementary Properties of Stationary Graphs. Let $X = (x_{ij})$ be a stationary graph.

9.1. $X \circ X \sim X$, that is,

$$x_{ij} = x_{k\ell} \iff \sum_s x_{is} x_{sj} = \sum_s x_{ks} x_{s\ell}.$$

9.2. $X \sim X'$, that is,

$$x_{ij} = x_{k\ell} \iff x_{ji} = x_{\ell k}.$$

9.3. $x_{ij} = x_{k\ell}$ implies $x_{ii} = x_{kk}$, $x_{jj} = x_{\ell\ell}$.

9.4. $x_{ii} = x_{kk}$ implies $\sum_s x_{si} = \sum_s x_{sk}$ and $\sum_s x_{is} = \sum_s x_{ks}$.

Proof. These properties are evident corollaries of 4.3, 4.4, 5, 6.1, 6.2.

Geometrically we interpret them as follows. If we have edges (say, x_{ij}, $x_{k\ell}$) of the same colour in X, then the sets of (ordered) paths of length 2 from the first vertices of these edges (i.e., i and k respectively) to the other vertices of these edges (i.e., j and ℓ respectively) contain the same number of paths of every colour. This is an interpretation of 9.1. The property 9.2 means that if two ordered pairs of vertices are connected by edges of the same colour, then the edges, connecting the same pairs of vertices but in opposite directions, also have the same colour.

The property 9.3 means that edges of the same colour are incident to equally coloured vertices.

The property 9.4 means that the set of colours (counted with multiplicity) of edges incident to equally coloured vertices is the same.

10. Stability with Respect to Paths of Greater Length.

Let $\langle i\text{-}j\rangle_t$ denote the set of the paths of length t from the vertex i to the vertex j of X. We say that $\langle i\text{-}j\rangle_t$ and $\langle k\text{-}\ell\rangle_t$ have the same composition if the multiplicities of the set of paths coloured in the same way are the same for both sets.

Theorem. Let $X = (x_{ij})$ be a stationary graph. If $x_{ij} = x_{k\ell}$, then $\langle i\text{-}j\rangle_t$ and $\langle k\text{-}\ell\rangle_t$ have the same composition.

Proof. By induction. For $t = 1$ there is nothing to prove. For $t = 2$, it is the definition of a stationary graph. Now remark (as already mentioned in the remarks after 4.4) that if Y and Z are graphs, then the (i,j)-th element of $Y \cdot Z$ describes paths of length 2 from i to j whose first edge belongs to Y and the second to Z. By induction $\langle i\text{-}j \rangle_{t-1}$ has the same composition as $\langle k\text{-}\ell \rangle_{t-1}$, that is, the (i,j)-th and (k,ℓ)-th elements of X^{t-1} coincide. But since X is stationary, the (i,j)-th and (k,ℓ)-th elements of $X^{t-1} \cdot X$ also coincide, as required.

11. The Matrix Algebra and the Basic Elements of a Stationary Graph.

Let X be a stationary graph. Consider the set $\mathfrak{A}(X)$ of the matrices $A = (a_{ij})$ (with entries in some ring) such that

$$a_{ij} = a_{k\ell} \quad \text{if } x_{ij} = x_{k\ell}.$$

11.1. Lemma. The set $\mathfrak{A}(X)$ is a matrix algebra, stable under transposition.

This is a direct corollary of 9.1, 9.2.

11.1.1. The matrix X is a generic point of the algebra $\mathfrak{A}(X)$ (cf. Section L).

11.2. Again let X be a stationary graph, $m = \dim X$. Let $x_1, \ldots, x_m$ be the distinct variables which are the entries of X. Let e_k be (0,1)-matrices obtained by substitution of $x_k = 1$, $x_i = 0$ for $i \neq k$, in X. The matrices e_i form a base of the algebra $\mathfrak{A}(X)$. We call them the basic elements of X and $\mathfrak{A}(X)$. Since they can be considered as adjacency matrices of graphs (directed or not) we sometimes call them the basic graphs. We have

$$X = \Sigma x_i e_i.$$

Let us point out some properties of the set of basic elements.

11.2.1. If e_i is a basic element, then so is e_i' (it follows from 9.2).

11.2.2. If e_i and e_j are basic elements, then

$$e_i e_j = \Sigma_k a_{ij}^k e_k$$

where a_{ij}^k are non-negative integers. (In fact, according to the remarks after 4.4, the number a_{ij}^k is the number of triangles of the form

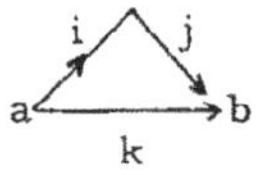

with fixed vertices a, b (connected by an edge of colour k)).

11.2.3. The graphs e_i are quasiregular in the sense that any vertex of this graph lying on an edge has the same number of entering edges and the same number of exiting edges (this follows from 9.4).

12. Examples.

a.

Γ : 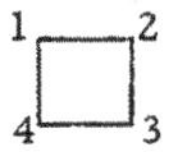

$$A(\Gamma) : \begin{pmatrix} 0 & 1 & 0 & 1 \\ 1 & 0 & 1 & 0 \\ 0 & 1 & 0 & 1 \\ 1 & 0 & 1 & 0 \end{pmatrix}, \quad X(\Gamma) = \begin{pmatrix} x & y & z & y \\ y & x & y & z \\ z & y & x & y \\ y & z & y & x \end{pmatrix}$$

$X \circ X = (p_{ij})$

$p_{11} = p_{22} = p_{33} = p_{44} = xx' + 2yy' + zz'$

$p_{12} = p_{14} = p_{21} = p_{23} = p_{32} = p_{34} = p_{41} = p_{43} = xy' + yx' + zy' + yz'$

$p_{13} = p_{24} = p_{31} = p_{42} = xz' + 2yy' + zx'$.

Thus, $X \sim \alpha(X)$, i.e., X is a stationary graph.

b.

Γ : 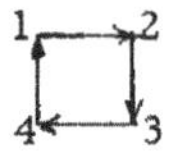

$$A(\Gamma) : \begin{pmatrix} 0 & 1 & 0 & 0 \\ 0 & 0 & 1 & 0 \\ 0 & 0 & 0 & 1 \\ 1 & 0 & 0 & 0 \end{pmatrix}, \quad X(\Gamma) : \begin{pmatrix} x & y & z & z \\ z & x & y & z \\ z & z & x & y \\ y & z & z & x \end{pmatrix}$$

$X \circ X = (p_{ij})$

$p_{11} = p_{22} = p_{33} = p_{44} = xx' + yz' + zz' + zy'$

$p_{12} = p_{23} = p_{34} = p_{41} = xy' + yx' + 2zz'$

$p_{13} = p_{24} = p_{31} = p_{42} = xz' + yy' + zx' + zz'$

$p_{14} = p_{21} = p_{32} = p_{43} = xz' + yz' + zy' + zx'$.

Thus, $X \circ X \sim Y = \begin{array}{cccc} x & y & z & u \\ u & x & y & z \\ z & u & x & y \\ y & z & u & x \end{array}$. It is easy to verify that $\alpha(Y) \sim Y$, hence $\hat{X} = Y$.

c.

Γ :

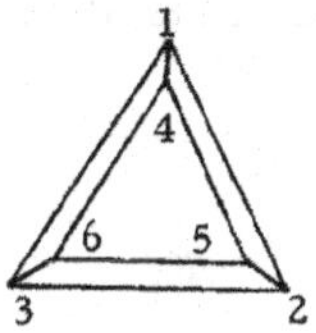

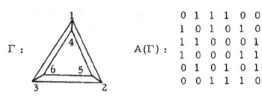

$A(\Gamma)$: $\begin{array}{cccccc} 0 & 1 & 1 & 1 & 0 & 0 \\ 1 & 0 & 1 & 0 & 1 & 0 \\ 1 & 1 & 0 & 0 & 0 & 1 \\ 1 & 0 & 0 & 0 & 1 & 1 \\ 0 & 1 & 0 & 1 & 0 & 1 \\ 0 & 0 & 1 & 1 & 1 & 0 \end{array}$

$$X(\Gamma) : \begin{array}{cccccc} x & y & y & y & z & z \\ y & x & y & z & y & z \\ y & y & x & z & z & y \\ y & z & z & x & y & y \\ z & y & z & y & x & y \\ z & z & y & y & y & x \end{array}$$

$X \circ X = (p_{ij})$

$p_{11} = p_{22} = \ldots = p_{66} = xx' + 3yy' + zz'$

$p_{15} = p_{16} = p_{24} = p_{26} = p_{34} = p_{35} = p_{42} = p_{43} = p_{51} = p_{53} = p_{61} = p_{62} =$

$= xz' + yz' + 2yy' + zy' + zx'$

$$X \circ X \sim Y = \begin{array}{cccccc} x & y & y & z & u & u \\ y & x & y & u & z & u \\ y & y & x & u & u & z \\ z & u & u & x & y & y \\ u & z & u & y & x & y \\ u & u & z & y & y & x \end{array}$$

It is easy to check that Y is stationary, $Y = \alpha(Y)$. Thus Y has exactly four basic elements:

e_1 (corresponds to x)

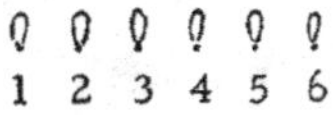

e_2 (corresponds to y)

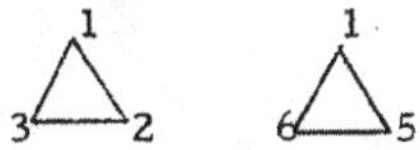

e_3 (corresponds to z) 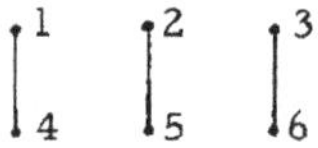

e_4 (corresponds to u) 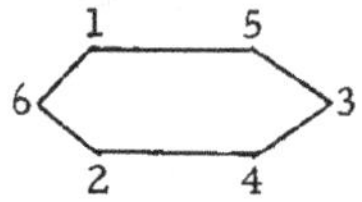

d. One more example of stabilization is given in the preceding section. Sections AA, AB, AC can also be considered as examples of stabilization.

D. PROPERTIES OF CELLS.

In the preceding section it was shown that the procedure of stabilization leads naturally to the stationary graph $\hat{X}$. It was shown also that a stationary graph $\hat{X}$ defines the matrix algebra $\mathfrak{A}(\hat{X})$. In this section we describe in detail the properties of a special class of such algebras. A more general (but also more formal) discussion of properties of algebras of this sort will be given in the next section. For examples, see Sections F, G.

The exposition below is based on [We 3]. The results are the analogues of certain well-known properties of permutation groups ([Wi 1], [Hi 2]).

1. We begin by giving an axiomatic definition.

1.1. Definitions. A cellular algebra is a matrix algebra $\mathfrak{A}$ having the following properties.

i) $\mathfrak{A}$ has a basis $B = \{e_i, i = 1,2,\ldots,d\}$, where the e_i are (0,1)-matrices. The basis $\{e_i\}$ is called a standard basis of $\mathfrak{A}$; standard bases differ only by the order of their elements.

ii) If $e_i \in B$, then $e_i' \in B$.

iii) $\Sigma e_i = I_n$, where n is the degree of matrices of $\mathfrak{A}$; n is called the degree of $\mathfrak{A}$.

iv) There exists an integer-valued function $d(e_i)$ such that the number of ones in any non-zero row of e_i is equal to $d(e_i)$. In this section (and also rather frequently elsewhere) we use the notation $n_i = d(e_i)$.

The basis $\{\xi_i\}$ of the underlying space V of the matrices of $\mathfrak{A}$ is called the standard basis of V.

The matrix $X = \Sigma x_i e_i$, where the x_i are independent variables is called the matrix of the cellular algebra $\mathfrak{A}$, written $X = X(\mathfrak{A})$. If $\mathfrak{A}$ is a cellular algebra with unity, then $X(\mathfrak{A})$ is a stationary graph. On the other hand, if X is a stationary graph, then there exists a cellular algebra $\mathfrak{A}$ with unity such that $X = X(\mathfrak{A})$; in this case we write $\mathfrak{A} = \mathfrak{A}(X)$.

1.2. Definition. A cellular algebra $\mathfrak{A}$ is called a cell if the number of ones in any row of every e_i is not zero.

1.3. Definition. A cellular algebra $\mathfrak{B}$ is called a cellular subalgebra of a cellular algebra $\mathfrak{A}$ if $\mathfrak{B}$ is a subalgebra of $\mathfrak{A}$.

In this case the elements of a standard basis of $\mathfrak{B}$ are sums of some elements of a standard basis of $\mathfrak{A}$. This follows from 1.1 i), iii).

2. Remark. The set of elements of the standard basis of a cellular algebra can be considered as a set of relations on $[1,n] \times [1,n]$. This set of relations forms a coherent configuration in the sense of D. G. Higman (cf. [Hi 3]). Conversely, any coherent configuration can be obtained in this way. So our cellular algebras and D. G. Higman's coherent configurations are equivalent objects. In [Hi 5], [Hi 6] D.G. Higman uses the term "adjacency algebra" where we use the term "cellular algebra".

3. Remarks. i) The application of 1.1 iv) to e_i' shows that the number of ones in any non-zero column of e_i is the same (and equal to $d(e_i')$).

ii) Geometrically, a cell with unity is a stationary graph X (cf., C 4.4) with the following properties:

a) All vertices of X have the same colour (are incident to loops of the same colour).

b) All basic elements of X are regular (cf. C 11.2.3).

4. Properties of Cells with Unity. Let $\mathfrak{A}$ be a cell with unity, B its standard

basis, $B = \{e_i,\ i = 0,1,\dots,d-1\}$, $e_0 = E_n$. Put $e_i e_j = \Sigma_k a_{ij}^k e_k$, $e_{i'} = e_i'$, $I = I_n$.

c1. $$\Sigma_s a_{ij}^s a_{sl}^k = \Sigma_s a_{is}^k a_{jl}^s .$$

Proof. This equality expresses the associativity of $\mathfrak{A}$. Geometrically, c1 has the following meaning. Consider the number $a_{ij\ell}^k$ of paths of length 3 and colour (i, j, ℓ) which are cut short by an edge of colour k. (By c10 this number does not depend on the edge of colour k but only on the sequence (i, j, ℓ, k).) This number can be computed in two ways. First, one can consider the paths of colour (s, ℓ) along the given edge, and for each $a_{s\ell}^k$ of such paths, one can consider the paths of colour (i, j) along its edge of colour s. The number of the latter is a_{ij}^s.

Thus for s fixed, the product $a_{s\ell}^k a_{ij}^s$ is equal to the number of paths of colour (i, j, ℓ) along an edge of colour k under the condition that the first and the third vertices are connected by an edge of colour s. Summing over s one obtains evidently the number $a_{ij\ell}^k$. Thus $a_{ij\ell}^k = \Sigma_s a_{ij}^s a_{s\ell}^k$.

On the other hand, one can consider paths of colour (i, s) along an edge of colour k and then paths of colour (j, ℓ) along an edge of colour s. In this case one obtains $a_{ij\ell}^k = \Sigma_s a_{is}^k a_{j\ell}^s$ (see figure above). Comparing these two expressions for $a_{ij\ell}^k$ we obtain our formula.

c2. $$n_{i'} = n_i .$$

Proof. In the case of a cell, the basic elements e_i and $e_{i'}$ are regular.

Therefore, for each vertex the entering and exiting valencies coincide.

c3. $$(\Sigma b_i e_i)\cdot I = I\cdot(\Sigma b_i e_i) = (\Sigma b_i n_i)I.$$

Proof. By 1.1 iv) and c2, $e_i \cdot I = I\cdot e_i = n_i I$. Hence our assertion follows by the distributive law of multiplication.

c4. $$\Sigma_i a^j_{ki} = \Sigma_i a^j_{ik} = n_k \text{ , } \Sigma n_k = n.$$

Proof. By c3, we have, $n_k I = I\cdot e_k = (\Sigma_i e_i)\cdot e_k = \Sigma_i (e_i e_k) = \Sigma_i \Sigma_s a^s_{ik} e_s = \Sigma_s \cdot (\Sigma_i a^s_{ik}) e_s = \Sigma_s n_k e_s$ whence $\Sigma_i a^s_{ik} = n_k$. Applying an analogous sequence of equalities to $e_k \cdot I$, we obtain $\Sigma_i a^s_{ki} = n_k$. The last equality is evident.

Geometrically, our property has the following interpretation. Consider a fixed edge of colour j and the paths of length 2 and colour (k, i) along it. If k, i are fixed, the number of these paths is a^j_{ki}. Summing over i we obtain the number of the paths of length 2 along the given edge, which begin with an edge of colour k, that is, we obtain the valency n_k of graph e_k.

c5. $$\Sigma_s a^s_{ij} n_s = n_i n_j.$$

Proof. $(n_i n_j)I = e_i \cdot (e_j \cdot I) = (e_i e_j)\cdot I = (\Sigma a^s_{ij} e_s)\cdot I = (\Sigma a^s_{ij} n_s)\cdot I.$

Geometrically, n_s is equal to the number of edges of colour s exiting from a vertex, and a^s_{ij} is equal to the number of paths of colour (i, j) along an edge of colour s. Hence, $\Sigma a^s_{ij} n_s$ is the number of all paths of colour (i, j) exiting from a fixed vertex. On the other hand, this number evidently equals $n_i n_j$, since n_i edges of colour i leave a vertex, and n_j edges of colour j leave the endpoints of all those edges.

c6. $a^0_{ij} = \delta_{i', j} n_i$; $a^i_{0j} = \delta_{ij}$ (where δ_{ij} is the Kronecker symbol).

Proof. Since $e_0 = E_n$ and $e_{i'} = e_i'$, we have $a_{ii'}^0 = n_i$. On the other hand, $\Sigma_j a_{ij}^0 = n_i$ by c4, hence $a_{ij}^0 = 0$ if $j \neq i'$. This proves the first equality. To prove the second, one considers the equality $e_j = e_0 \cdot e_j = \Sigma_i a_{0j}^i e_i$.

Geometrically, a_{ij}^0 equals the number of the paths of the form (i, j) whose source- and end-points coincide. Evidently, this number is zero if $i' \neq j$, and it is equal to the degree of e_i if $i = j'$. To interpret the second equality, let us consider those paths of length 2 along an edge of colour i which begin with the loop (colour 0) and then continue by an edge of colour j. It is evident that the number a_{0j}^i of those paths is 1 if $i = j$, and is 0 otherwise.

c7.
$$a_{ij}^k = a_{j'i'}^{k'} .$$

Proof. $\Sigma a_{ij}^s e_s' = (e_i e_j)' = e_j' e_i' = \Sigma a_{j'i'}^{s'} e_s'$. Geometrically, it is sufficient to consider, together with the paths (i, j) along an edge of colour k, the same paths in reverse direction. These latter are paths of the form (j', i') along the edge of colour k'.

c8.
$$n_i a_{jk}^{i'} = n_j a_{ki}^{j'} = n_k a_{ij}^{k'} .$$

Proof. Take c1 with $\ell = 0$:

$$\Sigma_s a_{jk}^s a_{si}^0 = \Sigma_s a_{js}^0 a_{ki}^s,$$

and use c6:

$$n_i a_{jk}^{i'} = \Sigma_s a_{jk}^s \delta_{si'} n_i = \Sigma_s a_{jk}^s a_{si}^0 = \Sigma_s a_{js}^0 a_{ki}^s = \Sigma_s \delta_{j's} n_j a_{ki}^s = n_j a_{ki}^{j'}.$$

The second equality is proved analogously.

Geometrically, equality c8 has the following meaning. Consider all paths of the form (j, k, i) whose source- and end-points coincide. First, fixing the third edge (of colour i) of one of these cycles, one sees that each such edge cuts short $a^{i'}_{jk}$ cycles (cf. Fig. below). Since from each vertex there exist n_i edges of colour i, one obtains $n_i a^{i'}_{jk}$ as the number of cycles of the

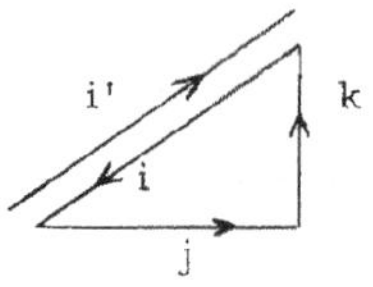

Fig.

form (j,k,i). On the other hand, fixing an edge of colour j one obtains j sets of cycles, and each set contains $a^{j'}_{ki}$ cycles.

c9. If $a^{i'}_{jk} \neq 0$, then $\frac{\max(n_i;n_j)}{(n_i,n_j)} \leq n_k$.

Proof. If $a^{i'}_{jk} \neq 0$, c8 implies $\frac{n_i}{n_j} = a^{i'}_{jk} = a^{j'}_{ik} \geq 1$. Therefore $a^{i'}_{jk}$ is a multiple of $\frac{n_j}{(n_i,n_j)}$. Since by c4, $n_k \geq a^{i'}_{jk}$, we obtain $n_k \geq \frac{n_j}{(n_i,n_j)}$. Similarly, $n_k \geq \frac{n_i}{(n_i,n_j)}$.

c10. The following assertions are proved similarly:

a) $a^{i'}_{kj}$ is a multiple of

$$\text{L.C.M.}\left(\frac{n_j}{(n_i,n_j)}, \frac{n_k}{(n_i,n_k)}\right) = \frac{n_j n_k (n_i,n_j,n_k)}{(n_i,n_j)(n_i,n_j)(n_j,n_k)}.$$

In addition, since $n_k \geq a^{i'}_{kj}$ we have:

b) $$n_k \geq \frac{(n_i,n_j)(n_i,n_k)(n_j,n_k)}{(n_i,n_j,n_k)} \quad \text{if } a^{i'}_{kj} \neq 0,$$

and analogous inequalities hold for n_i and n_j.

E. PROPERTIES OF CELLULAR ALGEBRAS OF RANK GREATER THAN ONE.

In this section we give short proofs of certain properties of general cellular algebras. These properties admit, of course, a similar geometrical interpretation as in the preceding section. After this we introduce some general notions (homomorphism, equivalence, etc.) and show how the corresponding properties can be established. The definition of a homomorphism requires reference to Sections H and I. This does not lead to a loop. The results in 5.2-5.5 are used in the study of correct cellular algebras (cf. J6) which play an important role in the algorithm of Section R. The assertion 5.6 is also used in this algorithm.

We begin with a proposition which shows that the matrix of a cellular algebra naturally falls into blocks ("cells"). This is the reason why our algebras are called cellular.

1. Decomposition of Cellular Algebras with Unity.

Proposition. Let $\mathfrak{A}$ be a cellular algebra with unity (i.e., $E_n \in \mathfrak{A}$, where $n = |\mathfrak{A}|$), and let $\{e_i\}_{i\in I}$ be its standard basis. Then

a) $E_n = \Sigma_{i\in I_0} e_i$, where I_0 is an appropriate set of indices;

b) $e_i^2 = e_i$, $e_i' = e_i$ for all $i \in I_0$;

c) let $T(i,j) = \{r : e_i e_r e_j = e_r\}$ for $i, j \in I_0$, then $I = \bigcup_{i,j\in I_0} T(i,j)$;

d) for any $i \in I_0$, $\mathfrak{A}_i = e_i \mathfrak{A} e_i$ is the cell with a standard basis $\{e_j, j \in T(i,i)\}$ with respect to the basis $\{\xi_t : e_i \xi_t = \xi_t\}$ of the space $V_i = e_i V$; e_i is the unity of $\mathfrak{A}_i$;

e) $\dim V_i = \mathrm{Sp}\, e_i$ for all $i \in I_0$.

Proof. Since $\{e_i\}$ is a basis of $\mathfrak{A}$ and $E_n \in \mathfrak{A}$, one has $E_n = \Sigma a_i e_i$. By D1.1 iii), the matrices e_i do not intersect. Thus $a_i = 0$ or 1. Setting $I_0 = \{i : a_i = 1\}$, one obtains a).

It is clear that the matrices e_i, $i \in I_0$, only have ones at the diagonal; this proves b).

Let us prove c). If $e_i e_r e_j \neq 0$, then $e = e_i e_r e_j \in \mathfrak{A}$, hence $e = \Sigma a_i e_i$. By b), it is clear that $e \subseteq e_r$. If $e \neq e_r$, then by D1.1 iii), the decomposition $e = \Sigma a_i e_i$ is impossible, hence $e = e_r$. This means that $e_i e_r e_j \neq 0$ implies $e_i e_r e_j = e_r$. Q.E.D.

The assertions d) and e) are now evident.

2. Corollary. By a simultaneous permutation of rows and columns the matrix $X = \Sigma x_i e_i$ can be brought into a block form:

$$X = (X_{ij})_{i,\, j\in I_0} = \begin{pmatrix} X_{11} & X_{12} \dots X_{1t} \\ \cdot & \cdot & \cdot \\ \cdot & \cdot & \cdot \\ \cdot & \cdot & \cdot \\ X_{t1} & X_{t2} \dots X_{tt} \end{pmatrix}$$

where $X_{ij} = \Sigma_{d\in T(i,j)} x_d e_d$.

If $N_i = \mathrm{Sp}\, e_i$, $i \in I_0$, then X_{ij} is an $(N_i \times N_j)$-matrix. A representation of X in the form (X_{ij}) will be called a central decomposition, or simply a decomposition of X.

Proof. Let $V_i = e_i V_i$, $i \in I_0$. From $\{\xi_i\} = \bigcup_t (\{\xi_j\} \cap V_i)$ and from 1d) our assertion follows.

3. Assumptions and Notations.

Let $\mathfrak{A}$ be a cellular algebra with unity, $X = X(\mathfrak{A})$ its matrix,

$X = (X_{ij})$ its decomposition (cf. 2 above). We write, and call

$|\mathfrak{A}| = |X|$ the degree of $\mathfrak{A}$ and X;

$\dim \mathfrak{A} = \dim X$ the dimension of $\mathfrak{A}$ and X;

$|I_0|$ the rank of $\mathfrak{A}$ and X;

X_{ij} the connection block of X_{ii} with X_{jj};

$V(\mathfrak{A}) = V(X)$ the vertices of X;

$V(X_{ii}) = \{j : e_i \xi_j = \xi_j\}$ the vertices of the cell X_{ii};

$e_{m'} = e'_m$ - it is clear that $m \in T(i, j)$ implies that $m' \in T(j, i)$;

$n_i = d(e_i)$ - the number of ones in any non-zero row of e_i;

$n_{i'} = d(e'_i)$ - the number of ones in any non-zero column of e_i:

$X = R$, $\mathfrak{A} = R - X$ and $\mathfrak{A}$ are split (i.e., $\dim X = |X|^2$, $\dim \mathfrak{A} = |\mathfrak{A}|^2$);

$X = S_n - X$ is the simplex, if $X = xE_n + y\tilde{I}_n$.

3.1. Remark. In his study of coherent configurations (cf. D2), D. G. Higman, uses the word "rank" where we use the word "dimension". Our "split" case corresponds to his "trivial".

4. General Properties of Structure Constants.

Proposition (compare D4). Put $N_\alpha = \mathrm{Sp}\, e_\alpha$ for $\alpha \in I_0$ and $\bar{e}_{\alpha\beta} = \sum_{t \in T(\alpha,\beta)} e_t$ for $\alpha, \beta \in I_0$. Let $\alpha, \beta, \lambda, \mu, \rho, \sigma \in I_0$, $m \in T(\alpha, \beta)$, $i \in T(\lambda, \mu)$, $j \in T(\rho, \sigma)$. Then

a) $n_{m'} \cdot N_\beta = n_m \cdot N_\alpha$;

b) $e_m \cdot e_i = 0$ if $\beta \neq \lambda$;

$$e_m \cdot e_i = \sum_{s \in T(\alpha, \mu)} a^s_{mi} e_s, \; a^s_{mi} \in \mathbb{Z}^+;$$

c) $\sum_s a^s_{ij} \cdot a^m_{sl} = \sum_s a^m_{is} \cdot a^s_{jl}$;

d) $\left(\sum_{m\in T(\alpha,\beta)} b_m e_m\right)\bar{e}_{\lambda\mu} = \left(\sum_{m\in T(\alpha,\beta)} n_m b_m\right)\delta_{\beta\lambda}\bar{e}_{\alpha\mu}$;

$\bar{e}_{\lambda\mu}\left(\sum_{m\in T(\alpha,\beta)} b_m c_m\right) = \left(\sum_{m\in T(\alpha,\beta)} n_{m'} b_m\right)\delta_{\mu\alpha}\bar{e}_{\lambda\beta}$;

e) $\sum_{j\in T(\beta,\sigma)} a_{ij}^m = \delta_{\mu\rho}\delta_{\sigma\beta}\delta_{\lambda\alpha} n_i$;

f) $a_{i'j'}^{m'} = a_{ji}^m$;

g) $\Sigma_s a_{mi}^s n_s = \delta_{\beta\lambda} n_m \cdot n_i$;

h) $a_{\rho i}^m = \delta_{im}\delta_{\rho\lambda}$ for all $\rho \in I_0$;

i) $a_{im}^\rho = \delta_{im'}\delta_{\rho\lambda}\cdot n_i$ for all $\rho \in I_0$;

j) $n_{m'} a_{ij}^{m'} = n_i a_{jm}^{i'}$.

<u>Proof.</u> a) is obtained by counting ones in e_m in two ways: $N\cdot n_{m'}$ (resp. $N\cdot n_m$) is the number of ones in any non-zero column (resp. row) multiplied by the number of non-zero columns (resp. rows).

b) If $\beta \neq \lambda$, then $e_\beta e_i = 0$, $e_m e_\beta = e_m$, hence $(e_m e_\beta)e_i = 0$. If $\beta = \lambda$, one has $e_\alpha \cdot (e_m e_i)\cdot e_\mu = (e_\alpha e_m)(e_i e_\mu) = e_m e_i$, whence $e_m e_i \subseteq \bar{e}_{\alpha\mu}$. It follows that

$$e_m e_i = \sum_{s\in T(\alpha,\mu)} a_{mi}^s e_s.$$

Since e_m, e_i, e_s are matrices with non-negative elements, from D1.1iii) one sees that $a_{ij}^s \geq 0$ and $a_{ij}^s \in \mathbb{Z}$.

c) Further, $e_i(e_j e_\ell) = (e_i e_j)e_\ell$, that is, $e_i(\Sigma_s a_{j\ell}^s e_s) = (\Sigma_s a_{ij}^s e_s)e_\ell$, whence $\sum_{s,t} a_{j\ell}^s a_{is}^t e_t = \sum_{s,t} a_{s\ell}^s e_t$. Comparing coefficients of e_t on both sides, one obtains c).

d) is evident.

e) Let us consider $e_i \bar{e}_{\rho\sigma} = n_i \delta_{\mu\rho}\bar{e}_{\lambda\sigma}$ (cf. d)). From

$\bar{e}_{\rho,\sigma} = \sum_{j\in T(\rho,\sigma)} e_j$ and $e_i\bar{e}_{\rho\sigma} = \sum_{s\in T(\alpha,\mu)} \sum_{j\in T(\rho,\sigma)} a_{ij}^{s} e_s$, one obtains e).

f) follows from the equality $(e_i e_j)' = e_j' e_i' = e_{j'} e_{i'}$.

g) One has $n_m n_i \delta_{\beta\lambda} \bar{e}_{\lambda\nu} = e_m(n_i \bar{e}_{\lambda\nu}) = e_m(e_i \bar{e}_{\mu\nu}) = (e_m e_i)\bar{e}_{\mu\nu} =$

$$= \Big(\sum_{s\in T(\alpha,\mu)} a_{mi}^{s} e_s\Big)\bar{e}_{\mu\nu} = \Big(\sum_{s\in T(\alpha,\mu)} a_{mi}^{s} n_s\Big)\bar{e}_{\alpha\nu}$$ whence g).

h) is the property 2 c) of the idempotents written in terms of a_{ij}^{k}.

i) becomes evident if we note that the diagonal entries of $e_i e_m$ are obtained by multiplication of t-th row of e_i by t-th column of e_m. By Dl. 1 ii), iii) the product is $\neq 0$ iff $m = i'$. In the latter case it is equal to the number of ones in the t-th row of e_i.

j) Let us use c) with $m = \beta$, $\ell = m$. We have $\sum_s a_{ij}^{s} a_{sm}^{\beta} = \sum_s a_{is}^{\beta} a_{jm}^{s}$. By i) $a_{sm}^{\beta} = \delta_{sm'} \cdot n_{m'}$, $a_{is}^{\beta} = \delta_{\beta\lambda} \cdot \delta_{si'} n_i$. Substituting this in our form of c) we obtain $a_{ij}^{m'} n_{m'} = \delta_{\beta\lambda} n_i a_{jm}^{i'}$. If $\beta \neq \lambda$, then $a_{jm}^{i'} = 0$ by b). This proves j).

5. <u>Weak Isomorphism, Homomorphism and Weak Equivalency.</u>

5.1. <u>Definitions.</u> Let X, Y be stationary graphs. We say X is <u>weakly isomorphic</u> to Y (written $X \underset{W}{\simeq} Y$) if $|X| = |Y|$ and there exists a substitution σ such that $\sigma X \sigma^{-1} \sim Y$; σ is called a weak isomorphism. We say that Y is a <u>homomorphic image</u> of $X = (X_{ij})$ if X_{ii} contains a normal subcell $\mathcal{L}_i$ (cf. Hl. 1) such that Y is weakly isomorphic to the factor-graph of X by the system $\{\mathcal{L}_i\}$ of normal subcells (cf. I4). We say that $X = \Sigma_{i\in I} x_i e_i$ and $Y = \Sigma_{j\in J} y_j f_j$ are weakly equivalent (written $X \simeq Y$) if there exists a one-to-one map $\varphi: I \to J$ such that $b_{\varphi(i)\varphi(j)}^{\varphi(k)} = a_{ij}^{k}$ where $e_i e_j = \Sigma a_{ij}^{k} e_k$, $f_i f_j = \Sigma b_{ij}^{k} e_k$. The map φ is called a <u>weak equivalency</u>. We sometimes write $\varphi: X \to Y$ and $\varphi(e_i) = f_{\varphi(i)}$ instead of $\varphi: I \to J$. We say that this mapping φ is a <u>natural</u> one if $\dim X = \dim Y$, $X = \Sigma_{i\in I} x_i e_i$, $Y = \Sigma_{i\in I} x_i f_i$ and $\varphi(e_i) = f_i$ (that is, the name of the variable is conserved under the mapping). A weak equivalency is called natural if the corresponding mapping of I onto J is natural. Of course, a natural weak isomorphism is an isomorphism.

5.2. <u>Proposition.</u> Let $X = (X_{ij})$ be a stationary graph, $X_{ij} = \Sigma_{i\in I} x_i e_i$. If $d(e_m) = d(e_{m'}) = 1$ for some $m \in J$, then $X_{ii} = X_{jj}$ and there exists a substitution $g \in \mathrm{Sym}(|X_{ii}|)$ such that $X_{ii} \sim X_{ij} \cdot g$.

<u>Proof.</u> Evidently one can assume that $\mathrm{rg}\, X = 2$, $i = 1$, $j = 2$. Since $d(e_m) = d(e_{m'}) = 1$ it follows from 4a) that $|X_{11}| = |X_{22}| = n$. Then $e_m|_{X_{12}}$ is the matrix of some substitution g. Let us substitute X_{12} by $X_{12}g^{-1}$, X_{21} by gX_{21}, X_{22} by $gX_{22}g^{-1}$ and X_{11} by the same X_{11}. Then $e_m|_{X_{12}}$ changes into E_n and X into an isomorphic graph. Since $\mathfrak{A}(X)$ is an algebra, we have $X_{11} \cdot X_{12} \subseteq X_{ij}$. Hence $X_{11} \cdot E_n \subseteq X_{12}$, that is, $X_{11} \subseteq X_{12}$. Analogously, $X_{22} \subseteq X_{12}$. On the other hand, $X_{21} \cdot X_{12} \subseteq X_{22}$ and $X_{12} \cdot X_{21} \subseteq X_{11}$ imply $X_{12} \subseteq X_{22}$ and $X_{12} \subseteq X_{11}$. Thus $X_{12} \sim X_{11} \sim X_{22}$. Q. E. D.

5.3. <u>Proposition.</u> Let $X = (X_{ij})_{i,j\in I}$ be a stationary graph. Let us write $i \sim j$ if there exists $e_m \subset X_{ij}$ such that $d(e_m) = d(e'_m) = 1$. Then this relation $\sim$ is an equivalency relation. In particular, $I = \cup I_t$ (a disjoint union) where $i, j \in I_t$ iff $i \sim j$.

<u>Proof.</u> Let $i \sim j$, $j \sim k$. Let us take $e_m \subset X_{ij}$, $e_n \subset X_{jk}$ with $d(e_m) = d(e_{m'}) = d(e_n) = d(e_{n'}) = 1$. Then $e_m e_n \subset X_{ik}$, and since $e_m|_{X_{ij}}$ and $e_n|_{X_{jk}}$ are permutation matrices, the matrix $e_m e_n|_{X_{ik}}$ is also a permutation matrix. Therefore, $e_m e_n = e_t$, $d(e_t) = d(e_{t'}) = 1$, that is, $i \sim k$. Since relations $i \sim i$ and

$i \sim j$ (iff $j \sim i$) are evident, we have proved our assertion.

5.4. Corollary. In notations of the preceding proposition, the stationary graph X is isomorphic to a stationary graph $X = (X_{ij})$ such that $X_{ij} \sim X_{mn}$ for all $i, j, m, n \in I_t$ and for any t.

Proof. Let $I_t = \{i_1, \ldots, i_r\}$. Choose from all $X_{i_1 j}$, $j \in I_t - i_1$, a matrix $e_{m(j)}$ with $d(e_{m(j)}) = d(e'_{m(j)}) = 1$. By permutation of $V(X_{jj})$ we bring $e_{m(j)}|_{X_{i_1 j}}$ into E_q, where $q = |X_{jj}|$. Then $c_{m(j)} e_{m(k)} = E_q \subset X_{jk}$ for $j, k \in I_t - i_1$. By 5.3 this implies that $X_{ij} \sim X_{mn}$ for all $i, j, m, n \in I_t$. Since the I_t do not intersect, the operation can be performed independently for all I_t, whence our assertion follows.

5.5. Proposition. If $\varphi : \mathfrak{A} \to \mathfrak{B}$ is a weak equivalency, then φ is an isomorphism of algebras, $\mathfrak{A}$ and $\mathfrak{B}$ have the same rank and degrees of a central decomposition (cf. 2).

Proof. Since, by definition, φ preserves the structure constants, φ clearly is an isomorphism of algebras.

From $d(e_i) = \Sigma_j a_{ij}^k$ (if this sum is not zero (cf. 4e)), it follows that $d(\varphi(e_i)) = d(e_i)$. This implies the remaining assertions.

5.6. Proposition. Let $\mathfrak{A}$ and $\mathfrak{B}$ be split cellular algebras. If $\varphi : \mathfrak{A} \to \mathfrak{B}$ is a natural weak equivalency, then φ is an isomorphism.

Proof. One has $n = |\mathfrak{A}| = |\mathfrak{B}|$, $\dim \mathfrak{A} = \dim \mathfrak{B} = n^2$. Let $E_n = \Sigma_{i \in I_0} e_i = \Sigma_{j \in J_0} f_j$, where $e_i \in \mathfrak{A}$, $f_j \in \mathfrak{B}$. By the preceding assertion, $\varphi(I_0) = J_0$ (since $i \in I_0$ iff $e_i^2 = e_i$). One can assume that $I_0 = J_0 = [1, n]$, $I = J = [1, n^2]$. After an appropriate permutation of the elements of a standard basis of the underlying space of

$\mathfrak{A}$ one can assume that

$$e_i = \mathrm{diag}(0,\dots,0,1,0,\dots,0)\ ,\ i \in I_0.$$

A permutation of the underlying space of $\mathfrak{B}$ brings f_i into the form

$$f_i = e_i \quad \text{for all} \quad i \in I_0 = J_0.$$

Assume now that the underlying spaces of $\mathfrak{A}$ and $\mathfrak{B}$ and their standard bases are identified and $e_i = f_i$ for all $i \in I_0$. I assert that in this united basis the equality $e_m = f_m$ holds for all $m \in I$. Indeed, if $m \in I$, there exists a unique pair $i, j \in I_0$, such that $e_i e_m e_j \neq 0$. Then since φ is a natural weak equivalency, we also have $f_i f_m f_j \neq 0$. From the above form of matrices $e_i = f_i$, $i \in I_0$, one concludes that $e_m = f_m$ for all $m \in I$ as asserted.

6. Some Numerical Invariants of Cellular Algebras. The numerical invariants introduced below are used in Section N.

6.1. Let $X = (X_{ij})_{i,j \in I}$ be a stationary graph. Let us assume that we are given a partition Π of the index set $I : I = \cup I_m$. If $p = (p_1, p_2, \dots, p_t)$ is a vector whose components p_i lie in some linearly ordered set, let us denote by p_{ord} the vector $(p_{i_1}, \dots, p_{i_t})$ with $p_{i_1} \geq p_{i_2} \geq \dots \geq p_{i_t}$. If $p = (p_i)_{i \in I}$ we denote by $p_{ord,\Pi}$ the vector whose components are ordered only within each I_m and components numbered by different I_m's donot mix. When we compare vectors of different length, we assume that short vectors are supplemented by zeros.

6.2. Let $X_{ij} = \sum_{T(i,j)} x_k e_k$. Put

$$\tilde{\mu}_1(X_{ij}) = (d(e_{i_1}), \dots, d(e_{i_r})),\ \text{where}\ \{i_1, \dots, i_r\} = T(i,j);$$

$$\mu_1(X_{ij}) = (|X_{ii}|, |X_{jj}|, \dim X_{jj}, (\tilde{\mu}_1(X_{ij}))_{ord});$$

$$\tilde{\mu}_2(X_{ii}) = (\mu_1(X_{ij}))_{j\in I};$$

$$\mu_{2,\Pi}(X_{ii}) = (|X_{ii}|, \dim X_{ii}, \mu_1(X_{ii}), (\tilde{\mu}_2(X_{ii}))_{ord,\Pi});$$

$$\tilde{\mu}_{3,\Pi}(X) = (\mu_{2,\Pi}(X_{ii})_{i\in I});$$

$$\mu_{3,\Pi}(X) = (\mathrm{rg}\, X, (\tilde{\mu}_{3,\Pi}(X))_{ord,\Pi}).$$

6.3. Remark. This set of invariants is sufficient for our immediate goals. Let us note, however, that one can consider iterations of these invariants. For example, one can substitute $|X_{ii}|$ in $\mu_1(X_{ij})$ by $\mu_{2,\Pi}(X_{ii})$, etc., until stabilization.

Furthermore, one can consider both matrix and vector invariants. For instance, one can consider the matrix $(\mu_{2,\Pi}(X_{ij}))_{i,j\in I}$. It is a matrix with linearly ordered entries. For this matrix one can construct its stationary graph (cf. M3) and subsequently compare such graphs. One can also put into correspondence with e_m the matrix (a^i_{mj}) and thereupon compare these matrices. One can substitute the latter matrices into the definition of $\mu_1(X_{ij})$. This leads to tensors of order 3, etc.

F. CELLULAR ALGEBRAS ARISING IN THE THEORY OF PERMUTATION GROUPS.

Although cellular algebras arise in different contexts (in our case as the result of graph stabilization) (cf. also [Hi 3]), the most important examples of cellular algebras are the centralizer rings of finite permutation groups (see below). It is known (the graphs of the 26-family of Section U are examples) that there exist cellular algebras which are not centralizers. However, centralizer rings provide us with a variety of notions and approaches which prove to be useful. Most of the constructions of Sections G-Q are based on the corresponding notions in permutation group theory.

1. Let G be a permutation group on a set $M = [1, n]$, and $M_1, \ldots, M_r$ be its orbits. Let us consider the vector space V with basis $\xi_1, \ldots, \xi_n$. We define an action of G on V by

$$\sigma\xi_i = \xi_{\sigma i}.$$

Let $\mathfrak{Z}(G, M)$ be the centralizer ring of G, that is, $\mathfrak{Z}(G, M) = \{B \in \mathfrak{M}_n : g^{-1}Bg = B \text{ for all } g \in G\}$. Then $\mathfrak{Z}(G, M)$ is a cellular algebra (cf. [Hi 2] and 2.3 below) with respect to the basis $\{\xi_i\}$, and $\{\xi_i\}$ is a standard basis of V. Below we shall construct a standard basis of $\mathfrak{Z}(G, M)$.

2. If $x \in M$, then G_x denotes the stabilizer of a point x. Let us choose a point x_i in each orbit M_i of G on M. Let $D_{1ij}, \ldots, D_{r(i,j),i,j}$ be orbits of G_{x_i} on M_j.

2.1. We construct graphs Γ_{mij} in the following manner. We connect the point gx_i with all points of the set gD_{mij}, for all $g \in G$. This definition does not depend on g. Specifically, if $gx_i = hx_i$, then $h^{-1}g \in G_{x_i}$, and therefore

$h^{-1}gD_{mij} = D_{mij}$, i.e., $gD_{mij} = hD_{mij}$. Let e_{mij} be the adjacency matrix of Γ_{mij}.

2.2. Assertion. The matrices e_{mij} form a basis of the centralizer algebra $\mathfrak{Z}(G, M)$.

Proof. We shall write $\mathfrak{Z}(G)$ for $\mathfrak{Z}(G, M)$. First, all matrices e_{mij} are in $\mathfrak{Z}(G)$. Actually, if $h \in G$, then in Γ_{mij} the incidence of hgx_i to points of hgD_{mij} implies that $h \in \operatorname{Aut}\Gamma_{mij}$. Hence e_{mij} commutes with all $h \in G$.

Take now $A \in \mathfrak{Z}(G)$. Since D_{mij} are orbits of G_{x_i}, the conditions $hx_i = x_i$; $hAh^{-1} = A$ imply that in row x_i of A all positions corresponding to D_{mij} are occupied by equal elements, say a_{mij}. Then the matrix $B = A - \sum_{m,i,j} a_{mij}e_{mij}$ is in $\mathfrak{Z}(G)$, and all elements of row x_i of B are equal to zero. Using the transitivity of G on M_i, and the condition $gBg^{-1} = B$, $g \in G$, one establishes that $B = 0$. Q.E.D.

2.3. Corollary. $\mathfrak{Z}(G, M)$ is a cellular algebra with unity and $\{e_{mij}\}$ is its standard basis.

Unity of $\mathfrak{Z}(G, M)$ is the sum of matrices $e_{m(i),i,i}$, where $D_{m(i),i,i} = x_i$.

2.4. Remark. There is no inverse correspondence, that is, in general $\mathfrak{Z}(G, M) = \mathfrak{Z}(H, M)$ does not imply $G = H$. As a trivial example, one can take the case $G = \operatorname{Sym} M$. Then for any doubly transitive group H on M one has $\mathfrak{Z}(G, M) = \mathfrak{Z}(H, M) = S_n$. A less trivial (but nonetheless very special) example is given in G2.6.

3. Remark. The purpose of our constructions is the following. If a graph Γ is given, how does one construct the matrix algebra $\mathfrak{Z}(\mathrm{Aut}\,\Gamma,\ V(\Gamma))$. Every stabilization (cf. Sections C, M, N, O) draws the algebras $\mathfrak{A}(\Gamma)$ and $\mathfrak{Z}(\mathrm{Aut}\,\Gamma, V(\Gamma))$ more and more together.

4. Let us once more point out the meaning of the notions of Sections D and E in the case when the cellular algebra is the centralizer algebra of some permutation group. Let G be a permutation group of a finite set M, $\widetilde{\mathfrak{A}} = \mathfrak{Z}(G, M)$, $\widetilde{X} = X(\widetilde{\mathfrak{A}})$.

4.1. Graphs e_m. If $\{e_i\}$ is a standard basis of $\widetilde{\mathfrak{A}}$, then G acts transitively on edges of every graph e_m. This explains our aspiration to find differences between non-diagonal variables of a graph.

4.2. If $X = (X_{ij})$ (cf. E2), then $V(X_{ii})$ is an analogue of an orbit of a group. In particular, for $\widetilde{X}$ the sets $V(\widetilde{X}_{ii})$ are orbits of G (and also of $\mathrm{Aut}\,\widetilde{X}$).

4.3. If $e_m \subset X_{ij}$, then $d(e_m)$ is an analogue of the length of the orbit of the stabilizer of $x \in V(X_{ii})$ on $V(X_{jj})$. For $\widetilde{X}$, this number is equal to the length of an orbit of G_x on $V(\widetilde{X}_{jj})$.

G. SOME CLASSES OF CELLULAR ALGEBRAS.

Some general classes of cellular algebras are constructed below. Most of them are modeled on the corresponding notions of permutation group theory or of the theory of algebras. Some of the classes introduced below are used in the canonization algorithm of Section R.

The description of the properties of some classes requires the use of results of subsequent sections. We introduce them now in the hope that they can be useful as a frame of reference.

1. Group Rings.

1.1. To each finite group G there corresponds the cell $\mathbb{Z}[G]$. It is defined in the following manner:

i) $V = \mathbb{Z}[G]$ is the group algebra of G, and the standard basis of V consists of elements of G;

ii) a standard basis of our cell consists of the operators of left multiplication $e_i = L_{g_i} : g \to g_i g$, for all $g \in G$.

We call this cell the group algebra (or ring) of G.

1.2. Clearly $d(e_i) = 1$, for every element of a standard basis of $\mathbb{Z}[G]$. The converse assertion is also true.

Proposition. Let $\mathfrak{A}$ be a cell. If $d(e_i) = 1$ for all e_i, then $\mathfrak{A} = \mathbb{Z}[G]$, where G is an appropriate group.

Proof. Since $d(e_i) = 1$, e_i is a permutation matrix. Since $\mathfrak{A}$ is an algebra, $e_i e_j = e_{m(i,j)}$. Therefore, matrices e_i form some group G. Clearly, $\mathfrak{A} = \mathbb{Z}[G]$.

1.3. Remark. $\mathbb{Z}[G]$ is the centralizer algebra (cf. the preceding section) of the permutation representation of G on itself by right translations.

2. Direct Sum.

2.1. Let $Y = (Y_{ij})_{i,j \in [1,m]}$, $Z = (Z_{ij})_{i,j \in [1,n]}$ be disjoint stationary graphs. Let us define the graph $X = Y \tilde{\oplus} Z = (X_{ij})_{i,j \in [1,m+n]}$ by the conditions:

$$X_{ij} = Y_{ij} \quad \text{for } i, j \in [1,m];$$

$$X_{i+m,j+m} = Z_{ij} \quad \text{for } i, j \in [1,n];$$

$$X_{ij} = \text{const} \quad \text{for } i \in [1,m],\ j \in [m+1, m+n]$$

$$\text{or for } i \in [m+1, m+n],\ j \in [1,m].$$

In addition, let the X_{ij} be all pairwise disjoint, and disjoint from Z and Y. The graph $Y \tilde{\oplus} Z$ will be called the direct sum of Z and Y. It is defined up to equivalence and depends only on the equivalence classes of Z and Y.

2.2. Proposition. $\operatorname{Aut} X = \operatorname{Aut} Y \times \operatorname{Aut} Z$ (direct product of permutation groups, cf. 2.6 below).

Proof. Aut X preserves $V(Y) \subset V(X)$ and $V(Z) \subset V(X)$ since Y and Z are disjoint. Take $g \in \operatorname{Sym} V(Y) \subset \operatorname{Sym} V(X)$, $g \in \operatorname{Aut} Y$. Then it follows from the constancy of the blocks X_{ij}, $i \in [1,m]$, $j \in [m+1, m+n]$, that $g \in \operatorname{Aut} X$. Q.E.D.

2.3. Proposition. If Y and Z are stationary graphs, then $X = Y \tilde{\oplus} Z$ is also stationary.

Proof. Evident.

2.4. Proposition. Let Y, Z be stationary graphs, $\mathfrak{A} = \mathfrak{A}(Y \tilde{\oplus} Z)$, $\mathfrak{B} = \mathfrak{A}(Y)$, $\mathfrak{C} = \mathfrak{A}(Z)$. Let $\overline{\mathfrak{A}}$, $\overline{\mathfrak{B}}$, $\overline{\mathfrak{C}}$ be ideals in the algebras $\mathfrak{A}, \mathfrak{B}, \mathfrak{C}$, respectively, complementary to the ideals $\mathfrak{A}, \mathfrak{B}, \mathfrak{C}$ of L6. Then $\overline{\mathfrak{A}} = \overline{\mathfrak{B}} \oplus \overline{\mathfrak{C}}$ (direct sum of algebras).

Proof. Evidently follows from definitions of the direct sum of graphs, and of the ideals $\tilde{\mathfrak{A}}, \tilde{\mathfrak{B}}, \tilde{\mathfrak{C}}$ (cf. L6).

2.5. Remark. This assertion shows that the notions of direct sum for algebras and for cellular algebras are rather close.

2.6. Remark. The direct sum of graphs is an analogue of the direct product of permutation groups. Namely, let G and H be permutation groups acting on V and W respectively. Let X and Y be the stationary graphs of the centralizer rings $\mathfrak{Z}(G, V)$ and $\mathfrak{Z}(H, W)$, respectively. Then $G \times H$ acts on $V \times W$, and $X \tilde{\oplus} Y$ is the stationary graph of $\mathfrak{Z}(G \times H, V \times W)$.

It is possible, however, for some group G acting on V, that the stationary graph of $\mathfrak{Z}(G, V)$ is a direct sum, but this action is not a direct product of actions of two different groups. This happens, e.g., if G has two orbits V_1 and V_2 on V, and G_x, $x \in V_1$, acts transitively on V_2 (since X_{12} = const by F4.3). A more concrete example: Take $G = Sym(5)$, $V_1 = Sym(5)/Sym(4)$, $V_2 = Sym(5)/N_{Sym(5)}Z_5$. Then $|G| = 120$, $|V_1| = 5$, $|V_2| = 6$. For $e_m \subset X_{12}$ we have by E4a): $5 \cdot d(e_m) = 6 \cdot d(e'_m)$. Therefore, 6 divides $d(e_m)$ and since $d(e_m) \leq 6$, we have $d(e_m) = 6$, whence X_{12} = const.

On the other hand, $Aut(X \tilde{\oplus} Y)$ is the direct product of Aut X and Aut Y.

3. Tensor Product.

3.1. Let Y, Z be graphs. Define the graph X in the following manner:

$$X = Y \tilde{\otimes} Z = (x_{ij,kl}),$$

where $x_{ij,kd} = x_{mn,rt}$ iff $y_{ik} = y_{mr}$ and $z_{jd} = z_{nt}$. $Y \tilde{\otimes} Z$ is called the <u>tensor product</u> of X and Y.

3.3. <u>Lemma.</u> If Y and Z are stationary graphs, then $Y \tilde{\otimes} Z$ is also a stationary graph and $\mathrm{rg}\, X = \mathrm{rg}\, Y \cdot \mathrm{rg}\, Z$.

<u>Proof.</u> Evident.

3.3. <u>Corollary.</u> If Y and Z are cells, then $X = Y \tilde{\otimes} Z$ is also a cell.

3.4. <u>Proposition.</u> If Y and Z are stationary graphs, $\mathcal{B} = \mathfrak{A}(Y)$, $\mathcal{L} = \mathfrak{A}(Z)$, $\mathfrak{A} = \mathfrak{A}(Y \tilde{\otimes} Z)$, then $\mathfrak{A} = \mathcal{B} \otimes \mathcal{L}$ (tensor product of algebras).

<u>Proof.</u> Follows evidently from the interpretation of X as a generic point of the algebra $\mathfrak{A}$.

3.5. <u>Proposition.</u> If Y and Z are cells, $X = Y \tilde{\otimes} Z$, then X contains two normal subcells $\mathcal{B}$ and $\mathcal{L}$ such that (cf. J5.4).

a) $X/\mathcal{B} = Y$, $X/\mathcal{L} = Z$;

b) $X(\mathcal{B}) \sim Z$, $X(\mathcal{L}) \sim Y$.

<u>Proof.</u> Let $|Y| \sim m$, $|Z| \sim n$. As a generic point of $\mathcal{B}$ (resp. $\mathcal{L}$) one can take the matrix $E_n \otimes Z$ (resp. $Y \otimes E_m$). Our assertions are now evident.

3.6. <u>Remark.</u> Theorem J5.4 shows that Proposition 3.5 is close to a characterization of tensor products.

3.7. <u>Remark.</u> A tensor product of graphs is analogous to both the tensor product of algebras and the direct product of permutation groups acting on the direct product of their domains. Explicitly, if G and H are permutation groups acting on V_1 and V_2, respectively, and if X and Y are stationary graphs corresponding to $\mathfrak{Z}(G, V_1)$ and $\mathfrak{Z}(G, V_2)$, respectively, then $X \tilde{\otimes} Y$ corresponds to $\mathfrak{Z}(G \times H, V_1 \times V_2)$.

4. Wreath Products. We shall define this construction only for cells.

4.1. Let $Y_1, Y_2, \ldots, Y_n$ be a set of naturally weakly equivalent cells (cf. E 5.1) of degree m, and let Z be a cell of degree n disjoint from Y_i. Let $Z = xE_n + \tilde{Z}$ where $\tilde{Z}$ has zero diagonal and entries different from those of $Y_1, \ldots, Y_n$. Put

$$X = (Y_1, \ldots, Y_n) \text{wr } Z = \Sigma h_i \otimes Y_i + \tilde{Z} \otimes I_n,$$

and let us call X the wreath product of the system $\{Y_i\}$ with Z. (Recall: h_i is the matrix with 1 only in position (i, i) and 0 in all other positions.)

4.2. Remark. A definition closer to that of group theory would be obtained in the case when $Y_1 = Y_2 = \ldots = Y_n$ and Y_i are cells (cf. 4.5 below). In this case the cell (cf. 4.3 below) X would be a subcell of the cell $Y_1 \tilde{\otimes} Z$.

4.3. Lemma. X is a cell.

Proof. Evident.

4.4. Lemma. If $\mathcal{L}$ is the normal subcell of X with the matrix $\Sigma h_i \otimes Y_i$, then $X/\mathcal{L} = Z$.

Proof. Evident.

4.5. Let $[1, n] = \cup I_t$ where Y_i is isomorphic to Y_j for all $i, j \in I_t$. Suppose that there exists no isomorphism of Y_i and Y_j if i and j do lie in the same I_t. Let G be the subgroup in Aut Z preserving all I_t, and G_t be the restriction of G to I_t. Let $H = \prod_i \mathrm{Aut}(h_i \otimes Y_i)$ (the direct product of permutation groups), $H_t = \mathrm{Aut}\, Y_i$, $i \in I_t$.

<u>Proposition.</u> $\mathrm{Aut}\, X = |H| \cdot |G|$; Aut X permutes the sets $V(h_i \otimes Y_i)$ in the same manner as G acts on I. The restriction of Aut X to $\bigcup_{i \in I_t} V(h_i \otimes Y_i)$ is $H_t \int G_t$ (the wreath product of H_t and G_t).

<u>Proof.</u> Clearly, Aut X preserves the partition of $V(X)$ into the sets $V(h_i \otimes Y_i)$. By 4.4 it permutes the sets $V(h_i \otimes Y_i)$ as some subgroup of Aut Z. Let us denote this subgroup by G. If there is no isomorphism of Y_i onto Y_j, then G cannot transfer i in j. Hence G preserves each I_t.

If $g \in \mathrm{Aut}\, X$, there exists a $g' \in G$ such that $g' g^{-1}$ preserves all $V(h_i \otimes Y_i)$ and induces an automorphism on each of them. Now our assertion follows immediately.

4.6. <u>Proposition</u> (compare L 7). The algebra $\mathfrak{A}(X)$ contains an ideal defined over $\mathbb{Q}$ and isomorphic as an algebra to the algebra $\mathfrak{A}(Z)$.

<u>Proof.</u> The subalgebra with a generic point X_C defined in H 7 for the normal subcell $\mathfrak{L}$ from Lemma 4.4 is, clearly, an ideal. The rest is evident.

4.7. <u>Example.</u> Let $Y_1, \ldots, Y_{15}$ be all distinct graphs of the 25-family (cf. Section U), and let X be the simplex, $Z = xE_n + y\tilde{I}_n$, where $x, y \notin Y_i$, $i \in [1,15]$. Then

$$X = (Y_1, \ldots, Y_{15}) \mathrm{wr}\, Z$$

is a cell (such a cell is called correct, cf. J6). The group Aut X preserves all subsets $[15j+1, 15(j+1)]$ for $j = 0, 1, \ldots, 14$, and it acts on $[15j+1, 15(j+1)]$ as the group Aut Y_{j+1}. Moreover, Aut X is the direct product of the groups Aut Y_j.

4.8. Although example 4.7 shows that the automorphism group of a wreath product of graphs can be very small, the notion of the wreath product of graphs is an analogue of the wreath product of groups. Indeed, let G and H be transitive permutation groups acting on V and W. Let us assume that $V = [1, n]$. Let X and Y be the stationary graphs of $\mathfrak{Z}(G, V)$ and $\mathfrak{Z}(H, W)$ respectively. Set $X_1 = X_2 = \ldots = X_n = X$. Then $(X_1, \ldots, X_n)$wr Y corresponds to $\mathfrak{Z}(G \int H, V \times W)$ (here $G \int H$ is the wreath product of G with H, cf. [Ha 2]).

H. IMPRIMITIVE CELLS AND CONSTRUCTION OF FACTOR-CELLS.

The notions introduced in this section are modeled on the corresponding notions of the permutation group theory. As in the group theory, they serve to reduce the study to the case of "primitive" cells (quotation marks can be omitted here). Probably, the passage to the quotient can be used in an algorithm of graph identification. Our use of this tool in the algorithm of Section R is indirect, and it relates only to correct stationary graphs.

The analogous situation in the permutation group theory is as follows.

Let G act transitively on V. If V is imprimitive, then $V = \bigcup_{i \in [1,n]} V_i$ and gV_i is some V_j for $g \in G$. This gives the action of G on $[1,n]$. In the terminology introduced below, $\mathfrak{z}(G,[1,n])$ is said to be the quotient of $\mathfrak{z}(G,V)$ by the normal subcell (cf. definition below) defined by $V = \bigcup_{i \in [1,n]} V_i$.

1. Let $\mathfrak{A}$ be a cell of degree n and $\{e_i\}$ be its standard basis.

1.1. Definitions. Let $\mathfrak{L}$ be a subcell of $\mathfrak{A}$, $f_0, f_1, \ldots, f_k$ be a standard basis of $\mathfrak{L}$. We call $\mathfrak{L}$ a normal subcell if

i) for every $i < k$ there exists j such that $f_i = e_j$;

ii) there exists a permutation of a standard basis of V which brings $C = \Sigma_{i<k} f_i$ into block-triangular form

$$\begin{pmatrix} C_1 & 0 & 0 & \ldots & 0 \\ * & C_2 & 0 & \ldots & 0 \\ * & * & * & & C_t \end{pmatrix}.$$

The property "$\mathfrak{L}$ is a normal subcell of $\mathfrak{A}$" is denoted by $\mathfrak{L} \triangleleft \mathfrak{A}$.

We shall show below (see Lemma 2) that C can be brought into block-diagonal form with diagonal blocks of the same degree, $C = E_r \otimes I_m$. We shall write $|\mathcal{L}| = d(C)$ and call $|\mathcal{L}|$ the degree of the normal subcell $\mathcal{L}$. The fact mentioned above also implies that $\Sigma_{i<k} x_i f_i$ is of the form $\Sigma_{i=1}^{r} h_i \otimes X_i$, where every X_i is the matrix of a cell and h_i is the diagonal matrix with the only non-zero entry equal to one at position (i, i). $\Sigma_{i<k} x_i f_i$ is called the matrix of the normal subcell $\mathcal{L}$, or its generic point. If all X_i are isomorphic, $X(\mathcal{L})$ is simply X_i.

Sometimes, if it does not lead to misunderstanding, we call $\{f_i\}_{i<k}$ a standard basis of the normal subcell $\mathcal{L}$. One can assume that $f_i = e_i$ for $i < k$. We sometimes write $e_{\mathcal{L}}$ for matrix C to emphasize its dependence on $\mathcal{L}$. We write $e_i \in \mathcal{L}$ if $e_i \cap C = e_i$. The matrices $e_i \in \mathcal{L}$ are called the elements of a standard basis of $\mathcal{L}$. We write $\mathcal{L} = 1$ if $C = E_n$, $n = |\mathfrak{A}|$.

We say that a normal subcell $\mathcal{B}$ is contained in a normal subcell $\mathcal{L}$ (notation $\mathcal{B} \subseteq \mathcal{L}$) if $e_{\mathcal{B}} \subseteq e_{\mathcal{L}}$. A normal subcell $\mathcal{B}$ is trivial if either $e_{\mathcal{B}} = E_n$ or $e_{\mathcal{B}} = I_n$.

A cell which contains no non-trivial normal subcell is called primitive; in the contrary case it is called imprimitive.

If $\mathcal{L}$ is a normal subcell of $\mathfrak{A}$ and, as above, $\Sigma_{i<k} x_i f_i = \Sigma_{i=1}^{r} h_i \otimes X_i$, we call the system of sets $V(h_i \otimes X_i)$ the system of imprimitivity in $\mathfrak{A}$, associated with $\mathcal{L}$. Each set $V(h_i \otimes X_i)$ is called a set of imprimitivity in $\mathfrak{A}$.

Henceforth, up to the end of the section we assume that k is fixed and $f_i = e_i$ for $i < k$.

1.2. Examples.

1.2.1. Let us first take the case of $\mathfrak{Z}(G, V)$, G an imprimitive transitive permutation group of V. Let $V = \bigcup_{i\in[1,n]} V_i$ be some imprimitive system for (G, V). Let G_i be the subgroup of G preserving V_i, and X_i be the stationary graph of $\mathfrak{Z}(G_i, V_i)$. Then $\Sigma h_i \otimes X_i$ is the matrix of a normal subcell in $\mathfrak{Z}(G, V)$. The corresponding imprimitivity system is, of course, $\{V_i\}$.

1.2.2. All graphs of C12 are imprimitive cells.

1.2.3. Simple examples of primitive cells yield oriented cycles with prime number of vertices.

1.2.4. Petersen's graph is an example of a primitive cell of degree 10. It has three basic graphs:

e_0 : loops, graph;

e_1 : 

e_2 : the complementary graph of e_1.

1.2.5. Examples of imprimitive cells are constructed in G3,4.

2. <u>Lemma.</u> Let $\mathcal{L}$ be a normal subcell of a cell $\mathfrak{A}$. If $C = \Sigma_{i<k} e_i$ (in the assumptions of 1.1) has a block-triangular form, then it can be brought into block-diagonal form, such that diagonal blocks contain no zeros and have pairwise equal degrees.

<u>Proof.</u> Let us consider f_k and f'_k. They have the following form:

$$f_k = \left(\begin{array}{c|c} * & \begin{matrix} 1 & 1 & \dots & 1 \\ \dots & & & \dots \\ 1 & 1 & \dots & 1 \end{matrix} \\ \hline 0 & \underbrace{\qquad A \qquad}_{d} \end{array}\right) \qquad f'_k = \left(\begin{array}{c|c} * & 0 \\ \hline \begin{matrix} 1 & 1 & \dots & 1 \\ \dots & & & \dots \\ 1 & 1 & \dots & 1 \end{matrix} & A' \end{array}\right)\Big\} d$$

By definition of a cell (cf. D1.2) each row of A contains d ones. Therefore, $A = I_d$. But since $f'_k \cap f_k \neq 0$, we have from D1.1 iii) that $f_k = f'_k$.

Set $C = \sum_{i<k} e_i$. Since $f_k = f'_k$, $C = I_n - f_k$ (where $n = |\mathfrak{a}|$), we have $C' = C$, whence the first assertion of our lemma. Let Γ be the graph whose adjacency matrix is $C - E_n$, then Γ is a simple graph. Let $\Gamma_1, \Gamma_2, \ldots, \Gamma_r$ be its connected components. By our assumption, $r \geq 2$. Let us renumber the vertices of Γ in such a way that C is brought to the form

$$\begin{matrix} C_1 & 0 & 0 & \ldots & 0 \\ 0 & C_2 & 0 & \ldots & 0 \\ 0 & 0 & C_3 & \ldots & 0 \\ \ldots & & \ldots & & \ldots \\ 0 & 0 & 0 & & C_r \end{matrix}$$

where $C_i - E_{|C_i|}$ is the adjacency matrix of Γ_i.

Let us show that $C_i = I_{|C_i|}$. We have $\Sigma^n_{m=1} C^m = \Sigma_{i \leq k} a_i f_i$. Clearly, $a_k = 0$. Since the Γ_i are connected, we have $\Sigma^n_{m=1} C_i^m \supset I_{|C_i|}$. Therefore, by D1.1 iii) we have $\Sigma_{i<k} f_i = \Sigma_{i<k} e_i = I_{|C_i|}$, whence $C_i = I_{|C_i|}$.

Since $|C_t| = d(C) = \Sigma_{i<k} n_i = m$ does not depend on t, we have $|C_t| = m$ for all t, i.e., $C_t = I_m$, whence the last assertion of the lemma.

3. Bringing C into a block-diagonal form, we define a partition of the matrix $X = \Sigma x_i e_i$ into $(m \times m)$-blocks X_{ij},

$$X = \begin{pmatrix} X_{11} & X_{12} & \ldots & X_{rr} \\ \ldots & \ldots & \ldots & \ldots \\ X_{r1} & X_{r2} & \ldots & X_{rr} \end{pmatrix}.$$

Proposition. Any two rows (and columns) of blocks of the matrix $X = (X_{ij})$ either have coinciding composition or have disjoint composition.

Proof. Let σ_1 be a row of block X_{ij} and σ_2 be a row of X_{kl}. Suppose that $\Sigma_{\sigma_1} x_{ij} \neq \Sigma_{\sigma_2} x_{ij}$, that is, that our rows have different composition. Then there evidently exists a variable x_q such that $\Sigma_{\sigma_1} x_{ij} = p_1 x_q + \dots$, $\Sigma_{\sigma_2} x_{ij} = p_2 x_q + \dots$ and $p_1 \neq p_2$. Consider then the product $e_q \cdot C$. All entries of the row σ_1 of this matrix are equal to p_1 and all entries of row σ_2 are equal to p_2. By definition of a cell, it follows from the above that σ_1 and σ_2 have no variables in common, that is, our assertion holds.

Let us note that the above proof also holds for two rows of one block.

4. Definition. Blocks X_{ij} and X_{kd} are called similar (notation $X_{ij} \approx X_{kd}$) if for every row (column) of X_{ij} there exists a row (column) of X_{kd} with equal composition and if the same holds for the pair X_{kd}, X_{ij}.

Proposition. All diagonal blocks are pairwise similar (in other words, all cells X_{ii} are pairwise naturally weakly equivalent). Every non-diagonal block is not similar to a diagonal one.

Proof. Both assertions follow from 2 and from the definition of a cell. Indeed, all variables x_i, $i < k$, are in diagonal blocks, and all other variables are outside diagonal blocks.

Proposition. Two non-similar blocks X_{ij} and X_{kl} have no variables in common. In other words, if e_q has ones in X_{ij}, then $e_q \cap X_{kl} = 0$.

Proof. Let $X_{ij} \not\approx X_{kl}$. Then, by definition, X_{ij} contains a row σ such that all rows of X_{kl} have different (from σ) composition. By Proposition 3, σ and X_{kl} have disjoint composition. For any column τ of X_{ij}, X_{kl} does not contain the element $\tau \cap \sigma$ of this column (since it is an element of σ). Therefore, the compositions of τ and X_{kl} are disjoint (by Proposition 3 applied to columns). Since τ is an arbitrary column of X_{ij}, our assertion is proved.

6. Proposition. Let $X_{ij} \approx X_{kl}$ be similar blocks of X. Then

$$\Sigma_{X_{ij}} x_{uv} = \Sigma_{X_{kl}} x_{uv},$$

that is, similar blocks have the same composition.

Proof. It is sufficient to show that X_{ij} and X_{kl} have an equal number of rows of any given composition. Let σ_1 be a row of X_{ij}, and suppose that X_{ij} (resp. X_{kl}) contains p_1 (resp. p_2) rows of the same composition as σ_1. Note that by Proposition 3 no variable of the row σ_1 lies outside those p_1 (resp. p_2) rows. Therefore, if $p_1 \neq p_2$, any two columns of X_{ij} and X_{kl} differ by their composition. This yields a contradiction to the condition $X_{ij} \approx X_{kl}$.

7. Definition. Let $\mathfrak{A}$ be a cell with unity, $X = \Sigma x_i e_i$ its graph, $\mathcal{C}$ a normal subcell of $\mathfrak{A}$, m the degree of diagonal blocks of C, $n = m \cdot r$. A factor-cell is a graph Z (and its algebra $\mathfrak{A}(Z)$) of degree r defined by the conditions:

$$z_{ij} = z_{st} \text{ iff } X_{ij} \approx X_{st}.$$

The notation is $Z = X/\mathcal{C}$ and $\mathfrak{A}(Z) = \mathfrak{A}/\mathcal{C}$. The factor-cell $\mathfrak{A}/\mathcal{C}$ is also called the quotient of $\mathfrak{A}$ by $\mathcal{C}$.

Theorem. The factor-cell $\mathfrak{A}/\mathcal{C}$ is a cell.

Proof. Let us consider the matrix $X_C \sim Z \otimes I_m$ of degree n. (It is obtained from X by changing each block X_{ij} into a constant block $z_k \cdot I_m$ in such a manner that non-similar blocks give rise to different variables.)

To begin with, let us note that in the product $X_C \cdot X_C$, in place of entries of some block X_{ij}, there arise equal polynomials. Further, let us show that in place of equal entries of matrix X_C in $X_C \cdot X_C$ there arise equal polynomials. Suppose it is false for blocks X_{st} and X_{pq}, $X_{st} \approx X_{pq}$. Set $g_{ij} = \sum_{e_m \cap X_{ij} \neq 0} e_m$. Then one has $g_{ij} \cdot g_{kd} \cap X_{st} = aI_m$, $g_{ij} \cdot g_{kd} \cap X_{pq} = bI_m$, $a \neq b$, for some i, j, k, d. By definition of a cell, this contradicts our assumption that $X_{pq} \approx X_{st}$. Q. E. D.

Thus we have just shown that X_C is a generic point of an algebra (i.e., $(X_C)_{ij} = (X_C)_{kd}$ implies $(X_C \cdot X_C)_{ij} = (X_C \cdot X_C)_{kd}$). Since $X_C \sim Z \otimes I_m$, we have shown that Z is a stationary graph. Since by Proposition 4 the diagonal entries of Z are equal, it follows that Z is a cell.

8.1. Remark. If $\mathcal{L}$ and $\mathcal{B}$ are cells, then there exists an imprimitive cell $\mathfrak{A}$ with a normal subcell $\tilde{\mathcal{B}}$ "isomorphic" to $\mathcal{B}$ (in the sense that $\Sigma_{i<k} x_i e_i = E_r \otimes X(\mathcal{B})$) such that $\mathfrak{A}/\tilde{\mathcal{B}} \underset{w}{\simeq} \mathcal{L}$. Examples of such cells $\mathfrak{A}$ are given in G3,4.

8.2. Remark. In the case $\mathfrak{A} = Z[G]$, normal subcells are subcells $Z[H]$, where H is a subgroup of G. Factorization corresponds to the construction of the algebra of double cosets $H\backslash G/H$, which is $\mathfrak{z}(G, G/H)$. The sense of factorization in the general case of an imprimitive group (G, V) was described in the introduction to this section.

9. A Geometrical Interpretation of the Notions of Subsections 4-7. Let us consider a cell with unity $\mathfrak{A}$ and its graph X. Let $\mathfrak{A}$ contain a normal subcell, and let $e_0 = E_n, e_1, \ldots, e_{k-1}$, be its basic elements chosen as in 1.1. Let us consider the matrix $A = \Sigma_{0<i<k} e_i$.

Lemma 2. A is the adjacency matrix of the disjoint union of r complete graphs $\Gamma_1, \ldots, \Gamma_r$ having the same number m of vertices. The set of the edges of the graph X which connect $a \in V(\Gamma_j)$ with vertices of Γ_i, is characterized by a row of the block X_{ij}.

If $a, b \in V(X)$, then Proposition 3 asserts that the sets of the edges from a to $V(\Gamma_i)$ and from b to $V(\Gamma_j)$ are either the same (as to colouring and multiplicity) or have no colour in common. (The cases $a = b$ and/or $\Gamma_i = \Gamma_j$ are not excluded.) Finally, Proposition 5 and 6 assert that for any i, j, k, d the sets of edges leading from $V(\Gamma_i)$ to $V(\Gamma_j)$ and from $V(\Gamma_k)$ to $V(\Gamma_d)$ are either the same or have no colour in common.

The factor graph is constructed in the following manner. All vertices of each Γ_i are identified, and edges connecting new vertices "inherit" the "colour" of the set of edges leading from $V(\Gamma_i)$ to $V(\Gamma_j)$.

10. Definition. An oriented graph Γ is called strongly connected if there is an oriented path from any of its vertices to any other vertex. If, however, for every two vertices a and b at least one (and, possibly, only one) path exists from a to b or from b to a, Γ is called connected.

11. Lemma. Let $\mathfrak{A}$ be a cell and $\{e_i\}$ its standard basis. Let $A = \Sigma_{i \in J} e_i$ be the sum of some basic graphs. Let Γ be the graph whose adjacency matrix is A. If Γ is connected, it is strongly connected.

Proof. Let $a \in V(\Gamma)$. Let $D(a)$ be the set of vertices which can be reached from a. Put

$$A_a = \{b \in V(\Gamma) : b \neq a, \ a \in D(b), \ b \notin D(a)\}$$

$$B_a = \{b \in V(\Gamma) : b \neq a, \ a \in D(b), \ b \in D(a)\}$$

$$C_a = \{b \in V(\Gamma) : b \neq a,\ a \notin D(b),\ b \in D(a)\}.$$

It is evident that $V(\Gamma) = a \cup A_a \cup B_a \cup C_a$.

Now, in the graph Γ, from a there exist paths only to the points of $B_a \cup C_a$. Consider $b \in A_a$. Then from b there exist paths to points of $B_a \cup C_a \cup a$ and, possibly, to some points of A_a. If $A_a \neq \phi$, then $D(a)$ and $D(b)$, $b \in A_a$, have unequal cardinality. This contradicts C10 and D3ii). Hence $A_a = \phi$.

Hence $B_a \cup C_a \cup a = V(\Gamma)$. Since by C10 this equality holds for every $a \in V(\Gamma)$ it follows that Γ is connected.

12. Proposition. A cell $\mathfrak{A}$ with unity is imprimitive iff it contains a disconnected basic graph e_i, $i > 0$.

Proof. If $\mathcal{L}$ is a normal subcell of $\mathfrak{A}$, then every $e_i \in \mathcal{L}$ is disconnected. Suppose now that e_1 is disconnected. Put

$$B = \Sigma_{m=1}^{n} e_1^m = \Sigma b_i e_i.$$

It is clear from geometrical considerations that the graphs e_i for which $b_i \neq 0$ are disconnected, and that the vertices of each of their connected components are contained in the vertices of some connected component of e_i. It may be assumed that $b_i \neq 0$ for $i = 0,1,2,\ldots,k-1$, and $b_i = 0$ for $i > k$. Then $\Sigma_{i<k} e_i = E_r \otimes I_m$ for some r and m. Therefore, $e_0, e_1, \ldots, e_{k-1}$ is a standard basis of some non-trivial normal subcell. Q. E. D.

13. Theorem. A cell $\mathfrak{A}$ is imprimitive iff there exists a (proper) subset K of indices such that matrix $B = \Sigma_{i \in k} e_i$ has two equal rows and $B \neq I_n$.

Proof. Let $\mathcal{L}$ be a normal subcell of $\mathfrak{A}$. Then f_k (in notations of 1.1) satisfies our conditions.

Conversly, let $B = \Sigma_{i \in K} e_i \neq I_n$ contain equal rows. Let $\sigma_1 = \sigma_2 = \ldots = \sigma_q$ be the first q rows of B and suppose that $\sigma_s \neq \sigma_1$ for all $s > 0$. (This is not restriction of generality since it can be obtained by simultaneous permutation of rows and columns.) Let $d(B) = d$. Consider the product $B \cdot B'$. All entries of the principal $(q \times q)$-minor M of the matrix $B \cdot B'$ are equal to d. Furthermore, all other entries of the first q rows are $< d$. By definition of a cell it follows that $B \cdot B' = \Sigma_{i \in K_1} de_i + \Sigma_{i \in K_2} a_i e_i$, where $a_i < d$ for $i \in K_2$. As follows from the previous discussions, from D1.1iii) and from Lemma 11, all e_i, $i \in K_1$, are disconnected. Thus our cell is imprimitive by Proposition 12.

I. CONSTRUCTION OF THE QUOTIENT IN THE CASE OF CELLULAR ALGEBRAS OF RANK GREATER THAN ONE.

Similarly to the definition of factor-cells (i.e., rank one case, see the preceding Section), a definition of the quotient can also be given for general cellular algebras. Such a definition is needed to complete the picture.
It is in this generality that the notion of the quotient may be used in the study of isomorphisms of graphs.

Since the extension to the general case of cellular algebras does not require any new ideas, we give in this Section exact definitions and omit proofs (with the exception of the proof of Lemma 2).

1. Let $X = (X_{ij})$ be a stationary graph. Let $\mathcal{L}_i$ be a normal subcell of X_{ii} and $V(X_{ii}) = \bigcup_{i=1}^{m_i} V_t^i$ be the corresponding partition of $V(X_{ii})$ into imprimitivity sets. We do not exclude the cases $|V_t^i| = 1$ for all t and $|V_1^i| = |X_{ii}|$. Set $N_i = |X_{ii}|$, $k_i = |\mathcal{L}_i|$, $C_i = e_{\mathcal{L}_i} = \sum_{e_m \in \mathcal{L}_i} e_m$. The partition of $V(X_{ii})$ into sets V_t^i induces the partition of matrices X_{ij} into $(k_i \times k_j)$-blocks $X_{ij,ts} = X(V_t^i, V_s^j)$.

<u>Definitions</u>.

1.1. Two rows (columns) $\sigma \subset X_{ij,ts}$ and $\tau \subset X_{ij,pq}$ are called <u>similar</u> (notation $\sigma \approx \tau$), if

$$\sum_{x_{uv} \in \sigma} x_{uv} = \sum_{x_{uv} \in \tau} x_{uv}$$

1.2. Two blocks $X_{ij,ts}$ and $X_{ij,pq}$ are called <u>similar</u> (notation $X_{ij,ts} \approx X_{ij,pq}$), if for every row (column) $\sigma \subset X_{ij,ts}$ there exists a similar row (column) $\tau \subset X_{ij,pq}$ and the same holds with the roles of $X_{ij,st}$ and $X_{ij,pq}$ interchanged.

2. <u>Lemma</u>. Let $\sigma_1 \subset X_{ij,pq}$ and $\sigma_2 \subset X_{ij,ts}$ be two rows (columns) of blocks of X_{ij}. If $\sigma_1 \not\approx \sigma_2$ then σ_1 and σ_2 have no variables in common.

The Proof will be given only for rows. Let $\Sigma_1 = \sum_{x_{uv}\in\sigma_1} x_{uv}$, $\Sigma_2 = \sum_{x_{uv}\in\sigma_2} x_{uv}$. Since $\sigma_1 \not\approx \sigma_2$ there exists a variable x_r such that $\Sigma_1 = a_1x_r + \dots$, $\Sigma_2 = a_2x_r + \dots$ and $a_1 \neq a_2$. Consider the matrix $e_r \cdot C_j$. This matrix has a_1 for each entry of row σ_1 and a_2 for each entry of row σ_2. By the definition of a stationary graph, this implies that σ_1 and σ_2 have no variables in common.

3. Proposition

3.1. If $X_{ij,pq} \not\approx X_{ij,ts}$ then these blocks have disjoint composition.

3.2. If $X_{ij,pq} \approx X_{ij,ts}$ then these blocks have equal composition, that is

$$\sum_{x_{uv}\in X_{ij,pq}} x_{uv} = \sum_{x_{uv}\in X_{ij,st}} x_{uv} .$$

The Proof is exactly the same as for cells (see H5, 6).

4. Definition. The factor graph $Y = (Y_{ij})$ of a graph $X = (X_{ij})_{i,j\in I}$ by a system $\{\mathcal{L}_i\}$ of normal subcells $\mathcal{L}_i \triangleleft X_i$ is defined in the following manner:

a) The vertices of Y are the sets V_t^i, $i \in I$, $t \in [1,m_i]$;

b) If $y_{pq}, y_{st} \in Y_{ij}$, then $y_{pq} = y_{st}$ if and only if $X_{ij,pq} \approx X_{ij,st}$.

5. Theorem. The factor-graph of a stationary graph $X = (X_{ij})$ by a system $\{\mathcal{L}_i\}$ of normal subcells $\mathcal{L}_i \triangleleft X_{ii}$ is a stationary graph of the same rank as X.

The Proof is the same as that of H7.

6. Notations. Keeping notations of 4, we write

$$Y_{ij} = \mathcal{L}_i \backslash X_{ij} / \mathcal{L}_j .$$

If $i = j$, we abbreviate

$$Y_{ii} = X_{ii}/\mathcal{L}_i .$$

If $\mathcal{L}_i = 1$ (i.e., $|V_t^i| = 1$ for all t) we write $Y_{ij} = X_{ij}/\mathcal{L}_j$, $Y_{ji} = \mathcal{L}_i \backslash X_{ij}$.

J. ON THE STRUCTURE OF CORRECT STATIONARY GRAPHS AND CELLS HAVING MORE THAN ONE NORMAL SUBCELL.

We begin this Section by stating simple properties of factorization and some conditions for existence of normal subcells.

We then pass to the study of cells with two or more normal subcells. The results here are analogous to results of Kuhn [Ku 1] about imprimitive permutation groups. We give these results to show how some of the notions introduced in the preceding Sections can be used to restrict the structure of stationary graphs. The annex, described below in 4.3, can be used in algorithms of graph canonization (although it is not used in Section R).

The Section is concluded by the study of correct cellular algebras. These algebras form an obvious obstruction to usual algorithms of graph canonization (cf. R 9.2). We describe in 6.7 a construction which permits dealing easily with these graphs. Other parts of Subsection 6 are dedicated to the proof that the approach based on 6.7 can be used and can be used with advantage (its use is described in R 5.4.2, R 6.2). A non-trivial example of a correct cell is given in G 4.7.

1. Elementary properties of factorization

Let X be a stationary graph and $X = (X_{ij})$ be its decomposition.

1.1. Proposition. Let $\mathcal{L}_i \triangleleft X_{ii}$, $\mathcal{L}_j \triangleleft X_{jj}$. Then

$$\mathcal{L}_i \backslash X_{ij} / \mathcal{L}_j = \mathcal{L}_i \backslash (X_{ij} / \mathcal{L}_j) = (\mathcal{L}_i \backslash X_{ij}) / \mathcal{L}_j .$$

Proof. Evident.

1.2. Proposition - Definition. Let $\mathcal{L} \triangleleft X_{ii}$, $\tilde{\mathcal{L}} \triangleleft X_{ii}/\mathcal{L}$. Then there exists $\mathcal{L}' \triangleleft X_{ii}$ such that $e_{\mathcal{L}} \subseteq e_{\mathcal{L}'}$, and the imprimitivity sets for $\mathcal{L}'$ are inverse images in $V(X_{ii})$ of the imprimitivity sets for $\tilde{\mathcal{L}}$ in $X_{ii}/\mathcal{L}$. $\mathcal{L}'$ is called the inverse image of $\tilde{\mathcal{L}} \triangleleft X_{ii}/\mathcal{L}$ in X_{ii}.

Proof. Evident.

1.3. Proposition. Let $\mathcal{L} \lhd X_{ii}$, $\mathcal{L}' \lhd X_{ii}$ and $e_{\mathcal{L}'} \subseteq e_{\mathcal{L}}$.

Then

a) $X_{ii}/\mathcal{L}'$ contains a normal subcell $\tilde{\mathcal{L}}$ whose inverse image in X_{ii} is $\mathcal{L}$;

b) $X_{ji}/\mathcal{L} = (X_{ji}/\mathcal{L}')/\tilde{\mathcal{L}}$.

Proof. Evident.

1.4. Corollary. A normal subcell $\mathcal{L}$ of X_{ii} is maximal (i.e., there does not exist $\mathcal{L}' \lhd X_{ii}$ with $e_{\mathcal{L}'} \subsetneq e_{\mathcal{L}}$) if and only if the cell $X_{ii}/\mathcal{L}$ is primitive.

The Proof follows immediately from 1.2.

2. A condition for the existence of normal subcells

Proposition (compare H13). Let $X = (X_{ij})$ be a stationary graph, $e_1, \ldots, e_r \subset X_{ij}$, $\bar{e} = \Sigma_{i=1}^{r} e_i$, $d(\bar{e}) < |X_{jj}|$. Assume that the non-zero rows of $\bar{e}$ with the indices $m_1, m_2, \ldots, m_t$ are equal, and that all other rows are not equal to these t rows. Then $\{m_1, \ldots, m_t\}$ is a set of imprimitivity for some normal subcell $\mathcal{L}$ of X_{ii}.

Proof. Put $W = \{m_1, \ldots, m_t\}$, and consider $\bar{e}\bar{e}' = (a_{pq})$. Then

$$a_{pq} = d(\bar{e}) \quad \text{if } p,q \in W ;$$

$$a_{pq} < d(\bar{e}) \quad \text{if } p \in W,\ q \notin W$$
$$\text{or } q \in W,\ p \notin W$$

Put $\bar{e}\bar{e}' = \Sigma a_i e_i$, $I = \{i : a_i = d(\bar{e})\}$. By the above discussion $I \neq \emptyset$, and the graph $f = (\Sigma_{r \in I} e_r)|_{X_{ii}}$ is disconnected. Moreover, $f \cap X(W, V(X_{ii})) = (I_t, 0)$. Thus our assertion follows.

Geometrically, our proof shows that if each vertex from a set of vertices W in $V(X_{ii})$ is connected by $\bar{e}$ in the same way to some set of vertices in $V(X_{jj})$, then this set W is distinguished. The fact that it defines an imprimitivity system follows from the properties of cells.

3. A condition for the existence of a homomorphism.

Proposition. Let $X = (X_{ij})$ be a stationary graph, $X_{ij} = \sum_{i \in I} x_i e_i$. If $d(e_{m'}) = 1$ for some $m \in I$, then X_{ii} is a homomorphic image of X_{jj}.

Proof. Put $d(e_m) = t$, $|X_{ii}| = n$,

$$
X \;=\; \left(\;
\begin{array}{c}
\boxed{\begin{array}{c} \overbrace{}^{V_1}\ \overbrace{}^{V_2}\ \overbrace{}^{V_3}\ \overbrace{}^{V_4} \\[4pt]
X_{ii} \qquad \underline{}\ \underline{} \qquad \qquad X_{ij} \\ \cdots \end{array}} \\[10pt]
X_{jj}
\end{array}
\;\right)
$$

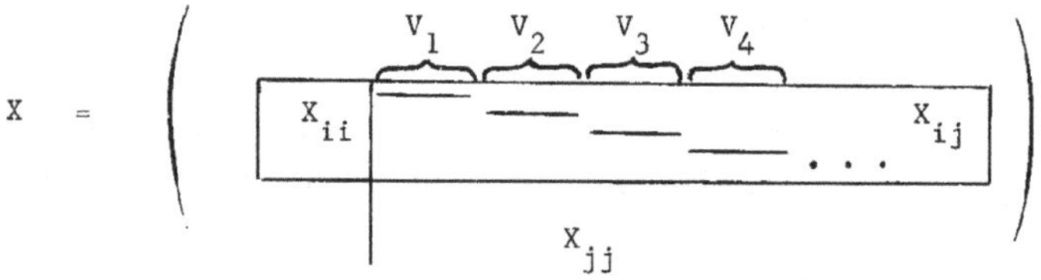

Since $d(e_{m'}) = 1$, $d(e_m) = t$, it follows that t columns of the matrix $e_m|_{X_{ij}}$ are pairwise equal, and that they are not equal to any other column of e_m. Let $\mathcal{D}$ be the normal subcell of X_{jj} defined (cf. Prop. 3) by equality of the rows of e_m. Passing to the quotient of X by the system $\{1 \triangleleft X_{pp},\ p \neq j,\ \mathcal{D} \triangleleft X_{jj}\}$ of normal subcells, we see that $X_{ij}/\mathcal{D}$ contains a basic element f_m (the image of e_m) which has the property $d(f_m) = d(f_{m'}) - 1$. Then by E 5.2, f_m defines a weak isomorphism of X_{ii} and $X_{jj}/\mathcal{D}$ as desired.

Geometrically, e_m serves to paint groups of vertices and edges of X_{jj} in the color of the vertices of $V(X_{ii})$. If equally painted vertices are identified, we get the graph X_{ii}.

4. Connection blocks of normal subcells.

4.1. Definition. Let X be a cell, and let $\mathcal{D}$ and $\mathcal{L}$ be two normal subcells of X. Let X_1, X_2, X_3, X_4 be four graphs which are equivalent to X and have pairwise disjoint composition. Then

$$
Y = \left(
\begin{array}{cc}
X_1 & X_2 \\[12pt]
X_3 & X_4
\end{array}
\right) = (Y_{ij})_{i,j \in [1,2]}
$$

is a stationary graph of rank 2. One takes $\mathcal{L} \triangleleft Y_{11}$, $\mathcal{L} \triangleleft Y_{22}$. Set

$$\mathrm{Con}_X(\mathcal{L}, \mathcal{L}) = \mathcal{L} \backslash Y_{12} / \mathcal{L},$$

and call this matrix the <u>connection block</u> of $\mathcal{L}$ with $\mathcal{L}$.

4.2. <u>Proposition.</u> Let X be a cell, $\mathcal{L} \triangleleft X$, $\mathcal{L} \triangleleft X$. Let $Z = (Z_{ij})_{i,j=1,2,3}$ be a stationary graph of rank 3 with

$$Z_{11} = X/\mathcal{L},\ Z_{22} = X/\mathcal{L},\ Z_{33} = X,\ Z_{13} = \mathrm{Con}_X(\mathcal{L},1),\ Z_{23} = \mathrm{Con}_X(\mathcal{L},1).$$

Then

$$Z_{12} = \mathrm{Con}_X(\mathcal{L}, \mathcal{L}).$$

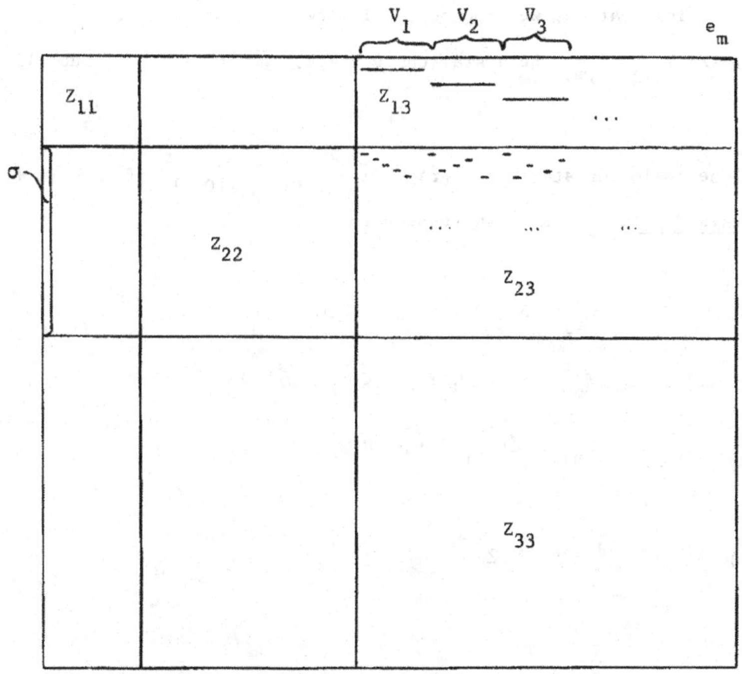

Proof. By definition of Z_{13} there exists a basic element $e_m \subset Z_{13}$ such that $d(e_m) = 1$, and e_m defines (cf. Prop. 3) a homomorphism of Z_{33} into Z_{11} (e_m is the image of the identity matrix, cf. Definition 4.1). It can be assumed that $e_m = E_r \otimes i_t$, where $r = |Z_{11}|$, $t = d(e_m)$. Let $\{V_1, V_2, \ldots, V_r\}$ be the imprimitivity system of $\mathcal{L}$. We can consider the similarity of those parts of the rows of the matrix Z_{23} which lie over sets V_i (see the definition of factorization, I4). Set $(a_{ij}) = e_m \cdot Z_{32} \subset Z_{12}$. It is clear that a_{ij}, $i \in V(Z_{11})$, $j \in V(Z_{22})$ depends only on the similarity class of the part of the row $X(j,V_i) \subset Z_{3,2}$. This means that $\mathrm{Con}_X(\mathcal{L}, \mathcal{L}) \subseteq Z_{12}$.

Now consider the first column $\sigma = X(V(Z_{22}),1)$ of the matrix Z_{21}. One has $\sigma \cdot e_m \subset Z_{23}$. Moreover, the first column of the matrix $\sigma \cdot e_m$ is equal to σ. Thus if the i-th and j-th elements of column σ are equal, then parts of rows $X(j,V_1)$ and $X(i,V_1)$ are similar, which yields the converse inclusion, i.e., $\mathrm{Con}_X(\mathcal{L}, \mathcal{L}) \supseteq Z_{12}$.

4.3. Annex.

Since the passage to the quotient simplifies a picture, but also leads to the loss of a useful information, we propose the following construction.

Let $X = (X_{ij})_{i,j \in [1,m]}$ be a stationary graph, $\mathcal{L}$ be a normal subcell of X_{tt}.

Definition. The following stationary graph $\tilde{X} = (\tilde{X}_{ij})_{i,j \in [0,m]}$ of rank $(m + 1)$ is called the annexed graph of X with respect to $\mathcal{L}$:

$$\tilde{X}_{ij} = X_{ij} \quad \text{if} \quad i,j \in [1,m]$$

$$\tilde{X}_{00} = X_{tt}/\mathcal{L}, \quad X_{0t} = \mathrm{Con}_{X_{tt}}(\mathcal{L}, 1)$$

$$\tilde{X}_{0j} = \mathcal{L} \setminus X_{tj} \quad \text{for} \quad j \neq t.$$

The block X_{00} is called the annex.

5. Imprimitive cells with several normal subcells

5.1. Let X be a cell, $\{V_i\}_{i\in[1,t]}$ and $\{\widetilde{V}_i\}_{i\in[1,\widetilde{t}]}$ be two systems of imprimitivity for X, and $\mathcal{L}$ and $\widetilde{\mathcal{L}}$ be the corresponding normal subcells of X.

Proposition. If $V_i \cap \widetilde{V}_j \neq \emptyset$, then X contains a normal subcell $\mathcal{L}'$ of degree $a = |V_i \cap \widetilde{V}_j|$ and $\mathcal{L}' \subseteq \mathcal{L}$, $\mathcal{L}' \subseteq \widetilde{\mathcal{L}}$.

Proof. Let $\{e_t,\ t \in J\}$ be the set of all those basic graphs e_t, whose edges connect points of the set $V_i \cap \widetilde{V}_j$. By the definition of normal subcells, and by Lemma H2, it is clear that $e_t \in \mathcal{L}$, $e_t \in \widetilde{\mathcal{L}}$ for all $t \in J$. Hence, any path along the edges of e_t must remain in both sets V_i and $\widetilde{V}_j$ (if it begins inside it). Consequently, $\Sigma_{t\in J}\, e_t$ defines a normal subcell $\mathcal{L}'$ and the set $V_i \cap \widetilde{V}_j$ is contained in some imprimitivity system of $\mathcal{L}'$. Q.E.D.

5.2. Corollary. Keeping the notation of 5.1 one has:

$$\text{if } |V_i \cap \widetilde{V}_j| = a \neq 0, \text{ then } a \,|\, (|\mathcal{L}|, |\widetilde{\mathcal{L}}|).$$

In particular, if $(|\mathcal{L}|, |\widetilde{\mathcal{L}}|) = 1$, then $|V_i \cap \widetilde{V}_j| = 1$ or 0 for all i,j.

5.3. Let us keep the notation of 5.1.

Proposition. Suppose that $\mathrm{Con}_X(\mathcal{L}, \widetilde{\mathcal{L}}) = \text{const}$. Then

a) There exists $a \in \mathbb{N}$ such that $V_i \cap \widetilde{V}_j = a$ for all i,j ;

b) $|V_i| = a\widetilde{t}$, $|\widetilde{V}_j| = at$, for all i,j, (recall that $t = |X/\mathcal{L}|$, $\widetilde{t} = |X/\widetilde{\mathcal{L}}|$),

c) There exists a normal subcell $\mathcal{L}' \triangleleft X$ such that $|\mathcal{L}'| = a$, $\mathcal{L}' \subseteq \mathcal{L}$, $\mathcal{L}' \subseteq \widetilde{\mathcal{L}}$.

Proof. Let $|V_1 \cap \widetilde{V}_1| = a$. Since $T = \mathrm{Con}_X(\mathcal{L}, \widetilde{\mathcal{L}}) = \text{const}$, and from the definition of connection blocks (cf. also 4.2), one concludes that $|V_1 \cap \widetilde{V}_i| = a$ for all i. This proves (a). Since $V_i \cap V_j = \emptyset$ and $V_1 \cap \bigcup_i \widetilde{V}_i = V_1$, one obtains (b) from (a). (c) coincides with 5.1.

5.4. Let $\mathcal{L}$ and $\tilde{\mathcal{L}}$ be two normal subcells of X, $f = e_{\mathcal{L}}$, $\tilde{f} = e_{\tilde{\mathcal{L}}}$, $t = |X/\mathcal{L}|$, $\tilde{t} = |X/\tilde{\mathcal{L}}|$.

Let us assume that $\mathrm{Con}_X(\mathcal{L}, \tilde{\mathcal{L}}) = \mathrm{const}$, and $|V_i \cap \tilde{V}_j| = 1$ for all i,j. (The latter assumption is not restrictive, since in the general case we can pass to the quotient of X by the normal subcell $\mathcal{L}'$ of 5.3.) We can assume that $V_i = [i(\tilde{t} - 1) + 1, i\tilde{t}]$ and $V_i \cap \tilde{V}_j = \{i(\tilde{t} - 1) + j\}$. Then

$$f = E_t \otimes I_{\tilde{t}}, \quad \tilde{f} = I_t \otimes E_{\tilde{t}}$$

Let $X = (X_{ij})$ (resp. $X = (\tilde{X}_{ij})$) be the partition of X into blocks corresponding to $\mathcal{L}$ (resp. to $\tilde{\mathcal{L}}$), $V(X_{ii}) = V_i$ (resp. $V(\tilde{X}_{ii}) = \tilde{V}_i$).

<u>Theorem</u>. (compare G3). In the above notations

a) $|\mathcal{L}| \cdot |\tilde{\mathcal{L}}| = |X|$

b) The cell $X|\mathcal{L}$ is weakly isomorphic to a subcell of all cells (X_{ii}), $i \in [1,t]$;

c) X_{ij} is embedded into the matrix $X_{ii} \vee X_{jj}$ (superimposition of X_{ii} and X_{jj}, confer C4.1).

<u>Proof</u>. (a) follows from 5.3b. Set

$$\bar{X} = \mathrm{diag}\,(X_{11}, X_{22}, \ldots, X_{tt})$$

Then from $f = I_t \otimes E_{\tilde{t}}$ and from $\bar{X} \cdot \tilde{f} \subseteq X$, $\tilde{f} \cdot \bar{X} \subseteq X$, the assertion of (c) follows.

Let us prove (b). Let $e_i \in \mathcal{O}\!\mathcal{L}(X)$. Define the $t \times t$-matrix $\bar{e}_i = (a_{pqi})$ in the following manner:

$$a_{pqi} = 1 \text{ if and only if } e_i \cap X_{pq} \neq 0,$$

$$a_{pqi} = 0 \text{ if and only if } e_i \cap X_{pq} = 0.$$

It is clear that $\bar{e}_i$ is a basic graph of the cell $X/\mathcal{L}$ (the image of e_i in $X/\mathcal{L}$). Identify $V(X/\mathcal{L})$ and $\tilde{V}_1$. Namely, $V_i \in V(X/\mathcal{L})$ is identified with $i(\tilde{t} - 1) + 1 \in \tilde{V}_1$). If e_i connects the points $p(\tilde{t} - 1) + 1$ and $q(\tilde{t} - 1) + 1$ of $\tilde{V}_1$, then $e_i \in \tilde{\mathcal{L}}$ and $\bar{e}_i$ connects blocks $V_p, V_q \in V(X/\mathcal{L})$. Hence $\bar{e}_i = e_i|\tilde{V}_1$. This yields the embedding of $X/\mathcal{L}$ into $\tilde{X}_{11}$. Since 1 can be replaced by any $r \in [1,\tilde{t}]$, without affecting the proof, our assertion is established.

5.5. <u>Corollary</u>. (compare [Ku 1]). Suppose that a cell X contains three normal subcells $\mathcal{L}_1, \mathcal{L}_2, \mathcal{L}_3$ and $\mathrm{Con}_X(\mathcal{L}_i, \mathcal{L}_j) = \mathrm{const}$ for all $i \neq j$. Suppose further that the cardinalities of the intersections of the imprimitivity sets of $\mathcal{L}_1$ and $\mathcal{L}_2$, and of $\mathcal{L}_2$ and $\mathcal{L}_3$, is one. Then

a) $|\mathcal{L}_1| = |\mathcal{L}_2| = |\mathcal{L}_3|$;

b) The cardinality of the intersections of the imprimitivity sets of $\mathcal{L}_1$ and $\mathcal{L}_3$ is also one.

<u>Proof</u> follows immediately from 5.3.

6. <u>Correct cellular algebras</u>

6.1. <u>Definitions</u>. Let $Y_1, \ldots, Y_n$ be arbitrary $(m \times r)$-matrices. The constant $(nm \times nr)$-matrices and the matrices (compare G4.1) of the form

$$X(Y_1, \ldots, Y_n) = \Sigma_{i=1}^{n} h_i \otimes Y_i + x\,\tilde{I}_n \otimes I_{m,r}$$

are called <u>fully correct</u>. (Recall that $h_i = \mathrm{diag}(0, \ldots, 0, 1, 0 \ldots, 0)$.) A stationary graph $X = (X_{ij})$ is called <u>correct</u> if there exist permutations σ_{ij} and τ_{ij} of degrees $|X_{ii}|$ and $|X_{jj}|$, respectively, such that for $i \neq j$ the matrices $\sigma_{ij} X_{ij} \tau_{ij}$ and for $i = j$ the matrices $\sigma_{ii} X_{ii} \sigma_{ii}^{-1}$ are fully correct.

6.2. <u>Proposition</u>. If X is a correct cell, then X contains a unique maximal normal subcell.

<u>Proof</u>. Let $\{e_0 = E_n, e_1, \ldots, e_r\}$ be a standard basis of X. One can assume

that $e_r = \tilde{I}_m \otimes I_t$, where $mt = n$. Then $\bar{e} = \Sigma_{i<r} e_i = I_n - e_r = E_m \otimes I_t$. Since the graph e_r is obviously connected, one concludes that for every normal subcell $\mathcal{L}$ in X, $e_{\mathcal{L}} \subseteq \bar{e}$. Thus $\bar{e}$ defines the unique maximal normal subcell. Q.E.D.

6.2.1. Remark. Geometrically, a correct cell is a stationary graph, whose vertices are partitioned, $V = \cup V_i$, and any pair of vertices from different V_i is connected by an edge of the same fixed color.

6.3. Corollary. Any factorcell of a correct cell is correct.

6.4. Theorem. Let X be a correct stationary graph. There exist permutations σ_i of degrees $|X_{ii}|$ such that $\sigma_i X_{ij} \sigma_j^{-1}$ is a fully correct matrix, for all i,j.

Proof. Let us bring the diagonal blocks X_{ii} into a fully correct form. Now consider non-diagonal blocks. Let $X_{ij} = \Sigma_{i=1}^{t} x_i e_i$. We can assume without loss of generality that $X_{ij} \neq$ const (in the contrary case X_{ij} is fully correct) and that $e_1 = \sigma(\tilde{I}_n \otimes I_{m,r})\tau$ for some permutations σ and τ. Then m rows of e_1 are pairwise equal. These rows define (Proposition 2) a normal subcell $\mathcal{L}$ of X_{ii}. Analogously, r equal columns of X_{ij} define a normal subcell $\mathcal{L}'$ of X_{jj}. Consider the factor-graph of X by this system of normal subcells. We have $\mathcal{L} \backslash X_{ij} / \mathcal{L}' = x f_1 + y f_2$, $d(f_2) = d(f_2') = 1$. Therefore (cf. E5.2) $X_{ii}/\mathcal{L} = S_n$, $X_{jj}/\mathcal{L}' = S_n$. Then 6.2 and 1.4 imply that $\mathcal{L}$ (resp. $\mathcal{L}'$) is the unique maximal normal subcell of X_{ii} (resp. X_{jj}).

Let $\mathcal{L}_i$ be the (unique) maximal normal subcell of X_{ii} (for every i). Consider the factor-graph $\bar{X}$ of the graph X by this system of normal subcells. As was shown above,

$$(*) \qquad \bar{X}_{ij} = \mathcal{L}_i \backslash X_{ij} / \mathcal{L}_j = x_{ij} f_{ij} + y_{ij} g_{ij}, \ d(g_{ij}) = d(g_{ij}') = 1$$

$$\text{if and only if } X_{ij} \neq \text{const},$$

$$(**) \qquad \mathcal{L}_i \backslash X_{ij} / \mathcal{L}_j = \text{const} \ \text{ if and only if } \ X_{ij} = \text{const}$$

Now using Corollary E5.4, one can bring $\bar{X}$ into a fully correct form. By (*)

and (**) the same permutation (mutatis mutandis) also brings X into a fully correct form.

6.5. Corollary. Every correct stationary graph is isomorphic to a fully correct graph $X = (X_{ij})$. Moreover, X is decomposable into the direct sum of fully correct graphs X_t having the following property:

Let $\mathcal{L}_i$ be the maximal normal subcell of X_{ii}. Then there exists a natural number $n = n(t)$, such that $X_{ij} \subset X_t$ implies that $\mathcal{L}_i \backslash X_{ij} / \mathcal{L}_j$ is of the form $xE_n + y\tilde{I}_n$.

Proof follows by 6.4(*), 6.4(**) from the proof of Theorem 6.4, from E5.4 and from 1.4.

6.6 Corollary. a) A cell $\Sigma_{i\in I} x_i e_i$ is correct if and only if there exists $e_m \subset X$ such that $e = \Sigma_{i\in I-m} e_i$ defines a normal subcell.

b) Let $X = (X_{ij})$ be a stationary graph, and let all cells X_{ii} be correct. Let $\mathcal{L}_i$ be the maximal normal subcells of X_{ii}. The graph X is correct if and only if, for every $X_{ij} \neq \text{const}$, there exists $e_m \subset X_{ij}$ such that

$$d(e_m) = |X_{jj}| - |\mathcal{L}_j|, \; d(e_{m'}) = |X_{ii}| - |\mathcal{L}_i|,$$

and such that $\mathcal{L}_i$ and $\mathcal{L}_j$ are defined by the equality of the non-zero rows and columns respectively of the matrix e_m.

6.7. Let $X = (X_{ij})_{i,j\in I}$ be a fully correct stationary graph, which cannot be decomposed into a direct sum of graphs (cf. 6.5). Then $X_{ij} \neq \text{const}$ for all i,j, and

$$X_{ij} = \Sigma_{k=1}^{n} h_k \otimes Y_{kij} + x\,\tilde{I}_n \otimes I_{m_i m_j} .$$

Definition. Put $Y_k = (Y_{kij})_{i,j\in I}$ and $F(X) = \{Y_1, \ldots, Y_n\}$. The set $F(X)$ is called the disassemblage of X.

<u>Lemma</u>. For all k, the graphs Y_k are stationary graphs of the same rank as X.

<u>Proof.</u> Evident.

6.8. <u>Theorem</u>. (compare G4.5). Let X, Y_k and $F(X)$ be as in 6.7. Suppose further that $[1,n] = \bigcup J_t$, and that Y_i, Y_j, $i, j \in J_t$, are isomorphic, but that Y_i and Y_j are not isomorphic when i, j come from different J_t. Then Aut X contains (as a subgroup of the group Sym $V(X)$) the direct product of permutation groups G_t which

a) preserve the partition of $V(X)$ into the sets $V(Y_{kii})$;

b) have the same induced action on J_t as Sym (J_t);

c) act on $\bigcup_{k \in J_t} V(Y_{kii})$ as the diagonal of some direct product of groups Sym (J_t).

<u>Proof</u>. Let $V_i = V(Y_i)$ and let $\{\xi_{1i}, \ldots, \xi_{ri}\}$ be those elements of the standard basis of the underlying space of X which lie in V_i. By our conditions, we can in addition assume (after an appropriate permutation of bases of V_i) that $Y_i = Y_j$ for $i,j \in J_t$. Then define the action of $\sigma \in \mathrm{Sym}(J_t)$ on $\{\xi_{mr}\}$ by

$$\sigma \xi_{m,r} = \xi_{m,\sigma(r)} \quad \text{for all } r \in J_t ,$$

$$\sigma \xi_{m,r} = \xi_{m,r} \quad \text{for all } r \notin J_t .$$

The σ defined in this manner commutes with the matrix $Z = (Z_{ij})$, where $Z_{ij} = \Sigma_{k=1}^{n} h_k \otimes Y_{kij}$. Since $X_{ij} - Z_{ij} = x\, \tilde{I}_n \otimes I_{m_i,m_j}$, σ also commutes with X, that is, $\sigma \in$ Aut X. Q.E.D.

K. PROPERTIES OF PRIMITIVE CELLS.

Sections H-J show the importance of factorization. The question arises: "What is the structure of those cells which cannot be factorized?" In particular, how can one describe the result of factorization? The properties of these cells (called primitive, cf. H 1.1) are mostly unknown. The results which we give below are of an arithmetical nature, that is, they give some restrictions on numbers a_{ij}^k. Such results possibly can be used to estimate the performance of an algorithm of graph identification.

The results of this Section are analogous to the results about primitive permutation groups (cf. [Wi 1, 17.5, 17.4, 18.7], [Hi 1, 4.1, 4.2]).

Let $\mathfrak{A}$ be a cell with unity, $\{e_i\}_{i\in I}$ its standard basis, $X = X(\mathfrak{A})$, $e_0 = E_n$. Suppose that $\mathfrak{A}$ is primitive (that is, does not contain nontrivial normal subcells, cf. H 1.1). Then all the basic graphs e_i, $i \in I-0$, are connected (by H 12).

1. Proposition. If $\mathfrak{A}$ is primitive and $n_i = 1$ for some $i \neq 0$, then $\mathfrak{A} \simeq Z[Z_p]$, the group algebra of the cyclic group Z_p, p a prime.

Remark. In this case, any basic graph e_i, $i \neq 0$, is an oriented cycle of length p.

To prove the Proposition we need the following:

1.1. Lemma. Let $\mathfrak{B}$ be a cell of degree n, and suppose that $n_0 = n_1 = \ldots = n_m = 1$, $m \neq 0$, and $n_i > 1$ for $i > m$. Then $e_0, e_1, \ldots, e_m$ define a normal subcell $\mathfrak{C}$ of $\mathfrak{B}$. They form a group of order m. In particular, m divides n.

Proof of the lemma. As in G1, $e_0,\ldots,e_m$ are permutation matrices. Since e_ie_j, $i, j \leq m$, is also a permutation matrix, we see that $e_ie_j = e_{k(i,j)}$ with $k(i,j) \leq m$. Therefore, the e_i, $i \leq m$, form a group of order m. Put $\bar{e} = \Sigma_{i\leq m} e_i$. Then $\bar{e}^2 = m\,\bar{e}$, whence it follows that $e_0, \ldots, e_m$ define a normal subcell.

1.2. Proof of Proposition 1. It follows from the assumptions of Proposition 1, and from Lemma 1.1, that $\mathfrak{A}$ contains a nontrivial normal subcell. Since $\mathfrak{A}$ is primitive, this normal subcell coincides with $\mathfrak{A}$. Therefore, $\mathfrak{A} \simeq \mathbf{Z}[G]$ for some group G. Since normal subcells of $\mathbf{Z}[G]$ correspond to subgroups of G (cf. H 8.2), it follows that G contains no proper subgroups. Therefore, $G \simeq Z_p$, p a prime.

Q.E.D.

2. Proposition. If a cell $\mathfrak{A}$ is primitive and $n_i = 2$ for some i, then $|\mathfrak{A}| = p$, p a prime, and all basic graphs e_i, $i > 0$, are non-oriented cycles of length p.

Remark. It can be expressed in the form $\mathfrak{A} \simeq \mathbf{Z}[\sigma + \sigma^{-1}]$, $\sigma^p = 1$, σ a permutation.

Proof. It follows from Lemma 1 and Proposition 1 that $n_i \geq 2$ for $i \neq 0$.

Now consider a basic matrix $e_i, n_i = 2$. One has $e_i e_i' = 2e_0 + e_k$, $d(e_k) = 2$ (since $2 = \min d(e_j)$), $e_k = e_k'$. By P. Hall's Theorem [Zy 1], e_k is representable as the sum of two permutation matrices

$$e_k = \sigma + \tau$$

Since $e_k = e_k'$, one has $\tau = \sigma^{-1}$. Hence e_k is the matrix of a non-oriented cycle. The cell $\mathfrak{A}(e_k)$ (cf. Section C) has the property that $d(f_i) = 2$ for any basic graph f_i of $\mathfrak{A}(e_k)$. Since $\mathfrak{A}(e_k)$ is a subcell of $\mathfrak{A}$, and since $q_1 = 2$, one has $\mathfrak{A} = \mathfrak{A}(e_k)$. If $|\mathfrak{A}| = m \cdot r$, $m, r \in \mathbf{N}$, one of the basic graphs is the disjoint union of m cycles of length r. Hence, if $|\mathfrak{A}|$ is not prime, $\mathfrak{A}$ is imprimitive, and our assertion is proved.

3. Lemma. If $n_i > 1$, $a_{ij}^k = n_j$, $k \neq 0$, then $n_i > n_j$.

Proof. We have $n_i = \Sigma_s a_{is}^k$ (by D 4 c 4), whence $n_i \geq n_j = a_{ij}^k$. Let us show that $n_i = n_j = a_{ij}^k$ contradicts the primitivity of $\mathfrak{A}$. By a simultaneous permutation of rows and columns, we can achieve that the first row of e_i and e_k take the form

$$(0 \;\ldots\; 0\; \underbrace{1 \;\ldots\; 1}_{n_i})$$

$$(0\; \underbrace{1 \;\ldots\; 1}_{n_k}\; 0 \;\ldots\; 0)$$

respectively. Consider now the first row of $e_i e_j$. By definition of a cell, it has the form

$$(* \underbrace{n_j \ldots n_j}_{n_k} * \ldots *)$$

It follows that, if $n_i = n_j$, all n_k columns of e_j with numbers $2, \ldots, (n_k + 1)$ have ones only in lower $n_i = n_j$ positions. Therefore, they are equal. However, if $n_k > 1$, it is impossible in an imprimitive cell (by H 13). The case $n_k = 1$ together with Proposition 1 contradicts our assumption (that $n_i > 1$). Therefore, the assumption $n_i = n_j$ is false. Q.E.D.

4. Let us order the numbers n_i. Let $q_1, \ldots, q_s$ be different values of n_i, $i \neq 0$. Suppose that $q_1 < q_2 < \ldots < q_m$. Set $I_k = \{i : n_i = q_k\}$.

Then $I\text{-}0 = \cup_k I_k$. We have $I_k \neq 0 \; \forall \, k$.

<u>Lemma</u>. For any two indices i,j there exists an $s \in I$ such that $s \neq j$, $a_{is}^j \neq 0$.

<u>Proof</u>. Since the graph e_i is connected, all entries of the matrix e_i^p are non-zero (where p is the maximum length of paths in e_i having no self-intersections, $p \leq n$). Thus, in the expression

$$e_i^p = \Sigma \, a_t \, e_t$$

all coefficients a_t are non-zero.

Let d be the least exponent such that

$$e_i^d = \Sigma \, b_t \, e_t, \; b_j \neq 0$$

Then, $e_i^{d-1} = \Sigma \, c_t \, e_t$, $c_j = 0$. We have

$$e_i(\Sigma \, c_t \, e_t) = \Sigma_{t,u} \, c_t \, a_{it}^u \, e_u = \Sigma \, b_t \, e_t, \; b_j \neq 0.$$

Hence, $\Sigma_t c_t a^j_{it} \neq 0$, that is, $a^j_{is} \neq 0$ for some s.

Geometrically, our assertion has the following meaning. No path of length $d - 1$ in the graph X consisting of edges of color i can be cut short by an edge of color j. However, such paths of length d exist. Consider such a path and consider the edge of some color s which cuts short the last $d - 1$ edges of this path. Then the edge of color s satisfies our requirements.

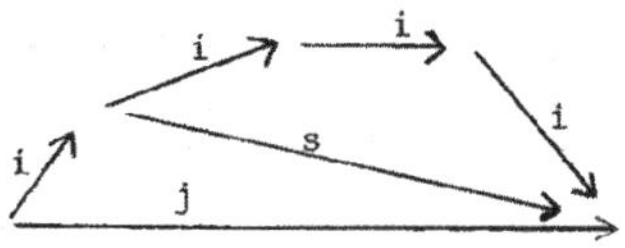

5. <u>Lemma</u>. For any I_t and any $i \notin I_t$, there exists $j \notin I_t$ and $k \in I_t$ such that $a^k_{ij} \neq 0$.

<u>Proof</u>. In the same manner as above, consider all possible paths consisting of edges of color i. There exists the minimal length d such that some of those paths are cut short by an edge of the color lying in I_t. Let an edge of color k, $k \in I_t$, cut short one of those paths, and let an edge of color j be the edge which cuts short the last $d - 1$ edges of this path. By the assumption of the minimality of d, one has $j \notin I_t$. Since $a^k_{ij} \neq 0$, we are done.

6. <u>Proposition</u>. $(q_s, q_m) \geq q_m q^{-1}_{m-1} > 1$ for all s.

<u>Proof</u>. Put $t = m$, take $i \in I_s$, and use the preceding assertion. There exist indices j and k such that $n_j \leq q_{m-1}$, $n_k = q_m$, $a^k_{ij} \neq 0$. From the inequality D 4 c 8 one gets

$$n_j \geq \frac{q_m}{(n_1, q_m)} ,$$

and this is our assertion.

7. <u>Proposition</u>. If $q_m = p^k$, p a prime, then $p^{k - \log_p q_{m-1}}$ divides all q_i.

8. <u>Proposition</u>. If $q_m = p$, p a prime, then $m = 1$, that is, all n_i are pairwise equal.

<u>Proof</u> of Propositions 7 and 8 is directly obtained by the application of Proposition 6.

9. $q_{k+1} \leq q_k q_1$ for all k; $q_m \leq q_1^{(\dim X)-2}$

<u>Proof</u>. Consider powers of the graph $f = \Sigma_{i \in I_1} e_i$. We have

$$f^t = \Sigma_i b_{it} e_i .$$

Put $I(t) = \{i : \exists s \leq t \ (b_{is} \neq 0)\}$, $q(t) = \max_{i \in I(t)} n_i$, that is, $q(t)$ is the greatest of the valencies of those graphs whose edges shortcut some paths having length $d \leq t$, and consisting of edges of colors from I_1. Evidently, $q(t+1) \leq q(t)q_1$. Since f is connected, one has either $I(t) = I$ or $I(t+1) \supset I(t)$ and $I(t+1) \neq I(t)$. Hence, $q_m = q(t_0)$, $t_0 \leq (\dim X) - |I_1 \cup 0| \leq (\dim X) - 2$.

L. ALGEBRAIC PROPERTIES OF CELLULAR ALGEBRAS.

We have associated with each graph a matrix algebra, $\mathfrak{A}(X)$. This gives rise to the question whether this algebra structure can be used to get new combinatorial information. In this Section we derive some information of this kind by purely algebraic methods. The results of this Section are analogous to some results about permutation groups. Most of them (and in a more general form) were also obtained by D. G. Higman [Hi 5],[Hi 6].

We assume some basic facts about the structure and representations of semi-simple associative algebras (cf., e.g., [Al 1]). Uninterested readers can skip this Section since its results are used in very few places.

If A is a matrix, we denote by A^* the conjugate matrix $\bar{A}'$ of the matrix A.

If $\mathfrak{A}$ is a cellular algebra and R is a $\mathbf{Z}$-ring, we denote the set of all R-rational points of the algebra $\mathfrak{A}$, by $\mathfrak{A}_R$, i.e., $\mathfrak{A}_R$ is the set of all those matrices which can be obtained by substituting elements of the ring R in place of variables in $X = X(\mathfrak{A})$. (In algebraic geometry, X would be called "the generic point of the matrix algebra $\mathfrak{A}$," and the matrices mentioned above would be called "the specializations of X.") Let $\mathfrak{M}_r$ denote the full matrix algebra of dimension r^2.

1. <u>Theorem</u>. Let $\mathfrak{A}$ be a cellular algebra with unity, and let $\{\xi_1,\dots,\xi_n\}$ be the standard basis of the underlying space V. Then

a) $\mathfrak{A}_{\mathbf{C}} \simeq \bigoplus_{i=1}^{t} \mathfrak{M}_{r_i}$ (as an algebra);

b) There exist $\mathfrak{A}_{\mathbf{C}}$-invariant and $\mathfrak{A}_{\mathbf{C}}$-irreducible subspaces $V_1, V_2, \dots, V_m$ of the space V, and bases $\zeta_{1i}, \dots, \zeta_{t_i i}$ of V_i ($m_i = \dim V_i$), such that $V = \bigoplus_{i=1}^{m} V_i$, and the matrix transforming the basis $\{\xi_i\}$ into $\{\zeta_{ij}\}$ is unitary.

<u>Proof</u> of this theorem is standard and uses the following

<u>Lemma</u>. Let V be a vector space over $\mathbf{C}$, $\mathfrak{A}$ be a set of linear operators on V, which contains with every operator A its conjugate A^*. Let $(u,v) = \Sigma u_i \bar{v}_i$ be the hermitian form on V. If W is an $\mathfrak{A}$-invariant subspace in V, then

$W^{\perp} = \{v \in V : (v,W) = 0\}$ is also an $\mathfrak{A}$-invariant subspace.

Proof of Lemma. Let $w \in W^{\perp}$, $A \in \mathfrak{A}$. Then $A^{*}W \subseteq W$ and $0 = (w,A^{*}W) = (Aw,W)$, that is $Aw \in W^{\perp}$. Q.E.D.

Proof of Theorem. Let W be an irreducible $\mathfrak{A}_{\mathbf{C}}$-invariant subspace in V. Let us put $W = V_1$. If $V_1, \ldots, V_d$ have already been constructed, we take for V_{d+1} any irreducible $\mathfrak{A}_{\mathbf{C}}$-invariant subspace in $(\bigoplus_{i=1}^{d} V_i)$. By the lemma above, this process leads to the construction of pairwise orthogonal subspaces $V_i \subseteq V$ and V is clearly the direct (even orthogonal) sum of these subspaces. We can now choose a basis $\{\zeta_{ij}\}$, $i \in [1, \dim V_j]$, in V_j, such that the vectors $\{\zeta_{ij}\}$ form an orthonormal basis of V. Then the matrix transforming the orthonormal basis $\{\xi_i\}$ into the orthonormal basis $\{\zeta_{ij}\}$ is a unitary matrix. This proves (b). Assertion (a) follows from (b) since any irreducible associative matrix algebra over $\mathbf{C}$ is isomorphic to some $\mathfrak{M}_r$.

2. Corollary. $\dim \mathfrak{A} = \sum_{i=1}^{t} r_i^2$.

3. Proposition. Let $\mathfrak{A}_{\mathbf{C}} = \bigoplus_{i=1}^{t} \mathfrak{N}_i$, $\mathfrak{N}_i \simeq \mathfrak{M}_{r_i}$. Let us denote by φ_i an isomorphism of $\mathfrak{N}_i$ on $\mathfrak{M}_{r_i}$, and by σ the involution $A \longrightarrow A^{*}$ of $\mathfrak{A}_{\mathbf{C}}$. Then the algebras $\mathfrak{N}_i$ can be renumbered so that

$$\mathfrak{N}_i^{\sigma} = \mathfrak{N}_{i+t_1} \qquad i \in [1, t_1] \;,$$

$$\mathfrak{N}_i^{\sigma} = \mathfrak{N}_{i-t_1} \qquad i \in [t_1 + 1, 2t_1] \;,$$

$$\mathfrak{N}_i^{\sigma} = \mathfrak{N}_i \qquad i \in [2t_1 + 1, t] \;.$$

Moreover, σ induces

on $\mathfrak{N}_i \oplus \mathfrak{N}_{i+t_1}$, $i \in [1, t_1]$, an involution of the second kind;

on $\mathfrak{N}_i$, $i \in [2t_1 + 1, t_2]$, an involution $A \longrightarrow \varphi_i^{-1}(S_i(\varphi_i(A))'S_i^{-1})$ with a symmetric matrix S_i; and

on $\mathfrak{N}_i$, $i \in [t_2 + 1, t]$, an involution $A \longrightarrow \varphi_i^{-1}(T_i(\varphi_i(A))'T_i^{-1})$ with a skew matrix T_i.

Proof. cf. [Wel].

4. Corollary. Algebras $\mathfrak{N}_i \oplus \mathfrak{N}_{i+t_1}$, $i \in [1, t_1]$, and $\mathfrak{N}_i$, $i \in [2t + 1, t]$ are defined over **R**.

If $i \in [2t_1 + 1, t]$, $r_i = 2m + 1$, then $\mathfrak{N}_i \simeq \mathfrak{M}_{r_i}$ over **R**.

If $i \in [2t_1 + 1, t]$, $r_i = 2m$, then over **R** either $\mathfrak{N}_i \simeq \mathfrak{M}_{2m}$ or $\mathfrak{N}_i \simeq \mathfrak{M}_m \otimes \mathfrak{L}$, where $\mathfrak{L}$ is the quaternion division algebra.

5. Corollary. If a is the number of symmetric basic elements e_i and $r = \dim \mathfrak{A}$, then $r - a = 2b$ for some $b \in \mathbf{Z}$ and $a + b = \sum_{i=1}^{t_1} r_i^2 + \sum_{i=2t_1+1}^{t_2} \frac{r_i(r_i+1)}{2} + \sum_{i=t_2+1}^{t} \frac{r_i(r_i-1)}{2}$

Proof. Let $2b$ be the number of e_i with $e_i' \neq e_i$. Then the equality $r = a + 2b$ is evident. Let us consider the second equality. Its right side gives (by Proposition 3) the dimension of the space of σ-symmetric elements in $\mathfrak{A}$. It is evident that its left side equals the same number.

6. Let $\mathfrak{A}$ be a cellular algebra of rank r with unity, and $X = X(\mathfrak{A}) = (X_{ij})$, $i, j \in [1, r]$, be its matrix. Let $\tilde{\mathfrak{A}}$ be the cellular algebra with a matrix $Y = (Y_{ij})$, $i, j \in [1, r]$, such that $V(Y_{ii}) = V(X_{ii})$, $Y_{ij} = \text{const}$ for all $i, j \in [1, r]$. (In particular, $\dim \tilde{\mathfrak{A}} = r^2$.) Put $N_i = X_{ii}$. Let us define the projection $\psi : \mathfrak{A} \longrightarrow \tilde{\mathfrak{A}}$ in the following manner: If $e_m \subset X_{ij}$, then $\psi(e_m)$ equals $\frac{d(e_m)E_{ij}}{N_j}$, where E_{ij} is the matrix with all ones in the block X_{ij} and zeros otherwise.

Proposition. a) $\tilde{\mathfrak{A}} \simeq \mathfrak{M}_r$ (over **Q**).

b) There exists a decomposition $\mathfrak{A} \simeq \tilde{\mathfrak{A}} \oplus \mathfrak{L}$ defined over **Q** where the injection $\varphi : \tilde{\mathfrak{A}} \longrightarrow \mathfrak{A}$ and the projection $\psi : \mathfrak{A} \longrightarrow \tilde{\mathfrak{A}}$ are defined as above.

c) $\tilde{\mathfrak{A}}_{\mathbf{Q}}$ acts as $\mathfrak{M}_r$ on some r-dimensional subspace W in V, and acts trivially on $W^{\perp}$.

Proof. (a) is evident. Let us prove (b). First, φ evidently is a ring-homomorphism. The projection ψ is also a ring-homomorphism since (cf. E4g)

$$d(e_m \cdot e_k) = \sum_s a_{mk}^s \, d(e_s) = \begin{cases} 0 & \text{if } e_m \subset X_{ij},\ e_k \subset X_{pq},\ j \neq p \\ d(e_m)\, d(e_k) & \text{if } e_m \subset X_{ij},\ e_k \subset X_{jq} \end{cases}$$

It remains to check that $\varphi\psi$ and $\psi\varphi$ are identities on $\tilde{\mathfrak{a}}$. This is evident.

Let us prove (c). Let $V = \bigoplus V_i$, $\dim V_i = N_i$, V_i corresponds to X_{ii}, cf. E1. Let $\xi_{1i}, \ldots, \xi_{N_i i}$ be the standard basis of V_i. Put $V_i^0 = \mathbf{C} \cdot (\sum_j \xi_{ji})$, $\overline{V}_i = \{v \in V_i,\ v = \sum v_j \cdot \xi_{ji},\ \sum v_j = 0\}$. Put further $V^0 = \bigoplus V_i^0$, $\overline{V} = \bigoplus \overline{V}_i$. Then it is clear that $\mathfrak{a}_{\mathbf{C}} V^0 \subseteq V^0$, $\overline{V} = (V^0)^{\perp}$, $\tilde{\mathfrak{a}}_{\mathbf{C}} \overline{V} = 0$. Since $V = \overline{V} \oplus V^0$, $\dim V^0 = r$, we are done.

7. Structure of an imprimitive cell

Let $\mathfrak{a}$ be a cell with unity, let $\mathfrak{L}$ be its normal subcell, let $e_0 = E_n, \ldots, e_r$ be a standard basis of $\mathfrak{L}$, and $\overline{e} = \sum_{i=0}^{r} e_i = E_d \otimes I_m$. Put $X = X(\mathfrak{a})$ and let $X = (X_{ij})$ be a partition of the X corresponding to $\mathfrak{L}$. The cell $\mathfrak{a}$ contains the subcell $\tilde{\mathfrak{a}}$ with matrix (cf. H7) $X_C = (Y_{ij})$, $V(Y_{ii}) = V(X_{ii})$, $Y_{ij} = \text{const}$ for all i,j, $Y_{ij} = Y_{st}$ if and only if $X_{ij} \approx X_{st}$. Clearly $\tilde{\mathfrak{a}}$ is an ideal in $\mathfrak{a}$.

Proposition. a) The subcell $\tilde{\mathfrak{a}}$ is isomorphic as an algebra to the factor-cell $\mathfrak{a}/\mathfrak{L}$ (the quotient is taken in the class of cells but not of associative algebras);

b) $f = \frac{1}{m}\overline{e}$ is an idempotent of $\mathfrak{a}$;

c) $f \cdot \mathfrak{a} \cdot f = \tilde{\mathfrak{a}}$;

d) d characteristic numbers of f are equal to 1 and $(m-1)d$ are equal to zero.

Proof. (a) coincides with H7. We have further, $\overline{e}^2 = m\,\overline{e}$, whence (b). If

$a \in f \cdot \mathfrak{A} \cdot f$, then $fa = af = a$. Hence in every block X_{ij} the rows and columns of of the matrix a are constant, i.e., $a \in \tilde{\mathfrak{A}}$. It is clear, that for $a \in \tilde{\mathfrak{A}}$ we have $af = fa = a$, i.e., $a \in f \cdot \mathfrak{A} \cdot f$. This proves (c). Property (d) follows from equality $\bar{e} = E_d \otimes I_m$.

8. Lemma. Let $\mathfrak{A}$ be a cellular algebra with unity, let $\{e_i\}$ be its standard basis. Let K be the field of quotients of a principal ideal domain R, $\mathfrak{A}_K = \Sigma Ke_i$, $\mathfrak{A}_R = \Sigma Re_i$. Let M be an $\mathfrak{A}_K$-module. Then M contains $\mathfrak{A}_R$-module M^* with $KM^* = M$. In other words, any K-representation of the algebra $\mathfrak{A}_K$ is K-equivalent to some R-representation.

Proof. Let $\eta_1, \ldots, \eta_t$ be a K-basis of M. Put $M^* = \sum_{i=1}^{\dim \mathfrak{A}} \sum_{j=1}^{\dim M} R\, e_i \eta_j$. Let $\eta_1^*, \ldots, \eta_t^*$ be an R-basis of the R-module M^* (remember, R is a principal ideal domain). The elements e_i are written in this basis as matrices with entries from R.

Q.E.D.

9. Let $\mathfrak{A}$ be a cell with unity, let $e_0 = E_n, e_1, \ldots, e_{r-1}$ be its standard basis, $n = |\mathfrak{A}|$, $r = \dim \mathfrak{A}$, $n_i = d(e_i)$. Put $\mathfrak{A}_{\mathbf{C}} = \bigoplus_{i=1}^{t} \mathfrak{N}_i$, $\mathfrak{N}_i \simeq \mathfrak{M}_{r_i}$ and let μ_i be the multiplicity of the nontrivial irreducible (over $\mathbf{C}$) representation of $\mathfrak{N}_i$ in the natural representation of $\mathfrak{A}_{\mathbf{C}}$ in V. Let $V = \bigoplus_{d=1}^{t} \bigoplus_{m=1}^{\mu_d} V_{d,m}$, where $V_{d,m}$ are $\mathfrak{A}_{\mathbf{C}}$-irreducible, $\mathfrak{N}_j V_{d,m} = \delta_{jd} V_{d,m}$ and $\mathfrak{A}_{\mathbf{C}}$-modules $V_{d,m}$ and $V_{d,q}$ are isomorphic for $m,q \in [1, \mu_d]$.

Theorem (Frame [Fr 1], [Wi 1, 30.5], [Hi 5]). In the above notations,

$$\frac{n^{r-2} \prod_{i=1}^{r-1} n_i}{\prod_{i=1}^{t} \frac{r_i^2}{\mu_i}} = q \in \mathbf{Z}$$

Moreover, $q = a \cdot \bar{a}$, where a is an algebraic integer.

Proof. By Proposition 1, there exists a unitary transformation U such that

$$U\, e_i\, U^{-1} = M_i = \bigoplus_{d=1}^{t} \bigoplus_{m=1}^{\mu_d} M_{d,m,i} \,,$$

where $M_{d,m,i}$ is $(r_d \times r_d)$-matrix of a linear transformation of the space $V_{d,m}$, and all matrices $M_{d,m,i}$, $m \in [1, \mu_d]$ determine equivalent irreducible representations of the algebra $\mathfrak{A}$. Note that

$$(U\, e_i\, U^{-1})' = \bar{U}\, e_i'\, \bar{U}^{-1} = \overline{(U\, e_i'\, U^{-1})} \,,$$

that is

$$(*) \qquad M'_{d,m,i'} = \bar{M}_{d,m,i'} \,.$$

Set

$$N_{ij} = Sp(M_{i'}\, M_j) \,.$$

We have by (D 4.c 6):

$$N_{ij} = Sp(e_{i'}\, e_j) = d(e_i) n\, \delta_{ij} \,.$$

On the other hand, using (*) and the decomposition of the matrix M_i, we have

$$\begin{aligned} N_{ij} &= Sp(M_{i'}\, M_j) = Sp(\bar{M}_i'\, M_j) \\ &= Sp(\bigoplus_d \bigoplus_m \bar{M}'_{d,m,i})(\bigoplus_d \bigoplus_m M_{d,m,j}) \\ &= \sum_d \sum_m Sp(\bar{M}'_{d,m,i}\, M_{d,m,j}) \end{aligned}$$

Now by the equivalency of representations, it follows that

$$Sp(\overline{M}'_{d,m,i}\, M_{d,m,j}) = Sp(\overline{M}'_{d,p,i}\, M_{d,p,j}) \quad \text{for all} \quad m,\ p \in [1,\ \mu_d]$$

Hence, setting $M_{d,i} = M_{d,1,i}$ we have

$$N_{ij} = \sum_d \mu_d\, Sp(\overline{M}'_{d,i}\, M_{d,j})$$

Suppose $M_{d,i} = (m^{\beta}_{\alpha,d,i})$, $\alpha,\ \beta \in [1,\ r_i]$. Then

$$(**) \qquad N_{ij} = \sum_{d=1}^{t} \mu_d \sum_{\alpha,\beta=1}^{r_d} \overline{m}^{\alpha}_{\beta,d,i} \cdot m^{\alpha}_{\beta,d,j} .$$

Let us number triads $(\alpha,\ \beta,\ d)$ where $\alpha,\ \beta \in [1,\ r_d]$, $d \in [1,\ t]$, $\sum_{i=1}^{t} r_i^2 = r$, with numbers of the interval $[1,\ r]$. Then

$$m^{\alpha}_{\beta,d,i} = a^{\sigma}_i,\ \sigma \in [1,\ r]$$

Let $R = \mathrm{diag}(\mu_1 E_{r_1^2},\ \mu_2 E_{r_2^2},\ \ldots .),\ A = (a^{\sigma}_i),\ N = (N_{ij})$. Then by (**)

$$N = \overline{A}'\ R\ A .$$

Let $a_1 = \det A$. Then

$$n^r \prod_{i=1}^{r-1} n_i = \det N = \det R \cdot a_1 \cdot \overline{a}_1 = (\Pi\, \mu_i^{r_i^2}) \cdot a_1 \cdot \overline{a}_1$$

Let us show that a_1 is an algebraic integer. By Lemma 8 in some basis $\eta_1,\ \ldots,\ \eta_{r_d}$ of $V_{d,1}$, the transformations e_i are written as matrices $T_{d,i}$ whose entries are algebraic integers. We have $T_{d,i} = S_d\, M_{d,i}\, S_d^{-1}$ where S_d is an appropriate matrix. Let us construct for matrices $T_{d,i}$ in the same manner, as above, the matrix $\widetilde{A}$. Since the linear transformation $F \longrightarrow S\ F\ S^{-1}$ of the space of matrices has determinant one, we have $\det A = \det \tilde{A}$. On the other hand, by the foregoing remark, $\det \tilde{A}$ is an algebraic integer. Hence a_1 is an algebraic integer.

It remains to show that a_1 is a multiple of n. Consider the sum $\sum_{i=0}^{r-1} a_i^{\sigma}$ for all rows of A. Let us assume that $\sigma = 1$ corresponds to the (one-dimensional and having multiplicity one) representation of $\mathfrak{A}$, described in 6. Then by (D4.c 4), $\sum_i a_i^{\sigma} = n\,\delta_1^{\sigma}$. Hence it follows that $a_1 = n \cdot a$, where a is the determinant of a $(r-1) \times (r-1)$-minor in A. Thus $a = a_1 \cdot n^{-1}$ is an algebraic integer. The theorem is proved.

Remark. If $\mathfrak{A}$ is commutative, then the entries of A are characteristic numbers of the basic elements e_i. Thus it is possible to obtain information on A and det A in this case.

10. Corollaries

a) If $\mathfrak{A}$ splits over $\mathbf{Q}$ into a direct sum of the full matrix algebras, then

$$q = a^2, \quad a \in \mathbf{Z}$$

b) If $\mathfrak{A}$ is commutative, and the characteristic numbers of all e_i are rational, then $q = a^2$, $a \in \mathbf{Z}$.

c) If $r_s = r_t$, $s \neq t$, implies $\mu_s \neq \mu_t$, and if $(r_i, \mu_i) = 1$ for all i, then

$$q = a^2, \quad a \in \mathbf{Z}$$

Proof. Under the conditions of (a), it follows from 8 that $a_1 = \det \tilde{A} \in \mathbf{Z}$ (in the notations of the proof of 9). Hence $a_1 = \bar{a}_1$, that is, $q = a^2$, $a \in \mathbf{Z}$. The conditions of (b) coincide with the conditions of (a) in the case of a commutative algebra $\mathfrak{A}$. Therefore, (b) follows from (a). Because of the first condition of (c), the simple summands of $\mathfrak{A}$ are defined over $\mathbf{Q}$. It follows that they are matrix algebras over simple division algebras. By the second condition of (c), those division algebras are fields whence we may use (a) again.

11. Let $\mathfrak{A}$ be a cell with unity. By 6, $\mathfrak{A} \simeq \mathfrak{M}_1 \oplus \mathfrak{L}$. We shall show that $\mathfrak{L}$ cannot be simple.

<u>Corollary</u>. The case $\mathfrak{A}$ is a cell, $\mathfrak{A} \simeq \mathfrak{M}_1 \oplus \mathfrak{M}_m$ (over $\mathbf{C}$), $m > 1$ is impossible. In particular, if $\dim \mathfrak{A} \leq 5$, then $\mathfrak{A}$ is commutative.

<u>Proof</u>. Let μ be the multiplicity of the irreducible representation of $\mathfrak{M}_m$, $m > 1$, in the natural representation of $\mathfrak{A}$. By 9

$$q = n^{r-2} \cdot \prod_{i=1}^{m^2} n_i \, \mu^{-m^2} \in \mathbf{Z}$$

By 1 and 6, we have $m \cdot \mu = n - 1$. In particular, $(\mu, n) = 1$. Thus $\prod_{i=1}^{m^2} n_i \, \mu^{-m^2} \in \mathbf{Z}$. However, $\sum_{i=1}^{m^2} n_i = n - 1$, and therefore $\prod_{i=1}^{m^2} n_i \leq \left(\frac{n-1}{m^2} \right)^{m^2}$ (since the maximum of the product is achieved by equality of multiples). We have in our case $(n - 1)m^{-2} = \mu m^{-1}$, i.e., $\prod n_i \leq (\mu m^{-1})^{m^2}$ whence it follows that $(\prod n_i)(\mu^{-m^2}) \leq m^{-m^2} < 1$ if $m > 1$. In particular, our q is not an integer. This contradiction proves our assertion.

12. <u>Rank 2</u>.

Let $\mathfrak{A}$ be a cellular algebra of rank 2 with unity

$$X = X(\mathfrak{A}) = \begin{pmatrix} X_{11} & X_{12} \\ X_{21} & X_{22} \end{pmatrix},$$

let $\mathfrak{A}'$, $\mathfrak{A}''$, $\mathfrak{A}_{12}$ and $\mathfrak{A}_{21}$ be the matrix algebras whose generic matrices are of the form

$$\begin{pmatrix} X_{11} & 0 \\ 0 & 0 \end{pmatrix}, \begin{pmatrix} 0 & 0 \\ 0 & X_{22} \end{pmatrix}, \begin{pmatrix} 0 & X_{12} \\ 0 & 0 \end{pmatrix}, \begin{pmatrix} 0 & 0 \\ X_{21} & 0 \end{pmatrix}$$

respectively. Let

$$\mathfrak{A}' = \oplus \mathfrak{N}'_i, \quad \mathfrak{N}'_i \simeq \mathfrak{M}_{r'_i}$$

$$\mathfrak{A}'' = \oplus \mathfrak{N}''_i, \quad \mathfrak{N}''_i \simeq \mathfrak{M}_{r''_i}$$

$$\mathfrak{A} = \oplus \mathfrak{N}_i, \quad \mathfrak{N}_i \simeq \mathfrak{M}_{r_i} .$$

Let μ_i (resp. μ'_i, μ''_i) be the multiplicity of the nontrivial irreducible representations of $\mathfrak{N}_i$ (resp. $\mathfrak{N}'_i$, $\mathfrak{N}''_i$) in the natural representation of $\mathfrak{A}$ (resp. $\mathfrak{A}'$, $\mathfrak{A}''$).

For any given $\mathfrak{N}_i$, let us define $I'(i) = \{j : \mathfrak{N}'_j \subset \mathfrak{N}_i\}$ and $I''(i) = \{j : \mathfrak{N}''_j \subset \mathfrak{N}_i\}$. Since $\mathfrak{A} \supset \mathfrak{A}' \oplus \mathfrak{A}''$ and $\mathfrak{N}'_j, \mathfrak{N}''_j$ are simple, it is clear that every $\mathfrak{N}'_j$ and $\mathfrak{N}''_j$ are contained in some $\mathfrak{N}_i$.

<u>Theorem</u>.

a) $\mathfrak{A}' \cdot \mathfrak{A}'' = 0$, $\mathfrak{A}' \cdot \mathfrak{A}_{12} \subseteq \mathfrak{A}_{12}$, $\mathfrak{A}' \cdot \mathfrak{A}_{21} = 0$, $\mathfrak{A}'' \cdot \mathfrak{A}_{12} = 0$, $\mathfrak{A}_{12} \cdot \mathfrak{A}_{12} = \mathfrak{A}_{21} \cdot \mathfrak{A}_{21} = 0$, $\mathfrak{A}_{12} \cdot \mathfrak{A}_{21} \subseteq \mathfrak{A}'$, $\mathfrak{A}_{21} \cdot \mathfrak{A}_{12} \subseteq \mathfrak{A}''$

b) $|I'(i)| \leq 1$, $|I''(i)| \leq 1$ for all i.

c) If $|I'(i)| = 0$, then $\mathfrak{N}_i \subset \mathfrak{A}''$. If $|I''(i)| = 0$, then $\mathfrak{N}_i \subset \mathfrak{A}'$.

d) If $|I'(i)| = 1$, $I'(i) = \{j\}$, then $\mu_i = \mu'_j$. If $|I''(i)| = 1$, $I''(i) = \{k\}$, then $\mu_i = \mu''_k$.

e) If $I'(i) = \{j\}$, $I''(i) = \{k\}$, then $r_i = r'_j + r''_k$ and $\mathfrak{N}'_j \oplus \mathfrak{N}''_k (\subset \mathfrak{N}_i)$ contains a maximal commutative semi-simple subalgebra of $\mathfrak{N}_i$.

<u>Proof</u>. Property (a) is verified directly. Let f_i (resp. f'_i, f''_i) be the central idempotent of $\mathfrak{N}_i$ (resp. of the algebras $\mathfrak{N}'_i$, $\mathfrak{N}''_i$ considered as the subalgebras in some $\mathfrak{N}_j$). Then $\mathfrak{N}_i = f_i \mathfrak{A} f_i$, $\mathfrak{N}'_i = f'_i \mathfrak{A} f'_i$, $\mathfrak{N}''_i = f''_i \mathfrak{A} f''_i$. By the definition of $I'(i)$ and $I''(i)$, one has

$$f_i = \sum_{j \in I'(i)} f'_j + \sum_{j \in I''(i)} f''_j = \tilde{f}'_i + \tilde{f}''_i .$$

Since $\mathfrak{N}_i \simeq \mathfrak{M}_{r_i}$, one has for $\tilde{f}'_i \neq 0$: $\tilde{f}'_i \mathfrak{N}_i \tilde{f}'_i \simeq \mathfrak{M}_t$ for some t.

It is clear by (a) that $\tilde{f}'_i \mathfrak{N}_i \tilde{f}'_i \subset \mathfrak{A}'$ and $\tilde{f}'_i \mathfrak{N}_i \tilde{f}'_i$ is a direct summand of $\mathfrak{A}'$. This implies that $\tilde{f}'_i \mathfrak{N}_i \tilde{f}'_i = \mathfrak{N}'_j$ for an appropriate j. Therefore, if $|I'(i)| = 0$, then $|I'(i)| = 1$, and (b) is proved. From this (c) and (e) follow. Let us prove (d). Since the multiplicity of the irreducible representation of $\tilde{f}'_i \mathfrak{N}_i \tilde{f}'_i$ (if $\neq 0$) in the irreducible representation of $\mathfrak{N}_i$ is equal to unity, we see that the multiplicity of the irreducible representation of $\tilde{f}'_i \mathfrak{N}_i \tilde{f}'_i$ in the μ_i-fold irreducible representation of $\mathfrak{N}_i$ is μ_i. This proves (d).

13. Corollaries. Let $X = (X_{ij})$ be a stationary graph.

a) $\dim X_{ij} \leq \frac{1}{2}(\dim X_{ii} + \dim X_{jj})$

b) If $X_{ii} = S_m$, $\dim X_{ij} = t + 1$, then $\mathfrak{A}(X_{jj})$ contains, as a direct summand, the subalgebra $\mathfrak{N}''_1 \oplus \mathfrak{N}''_2$, where $\mathfrak{N}''_1 \simeq \mathfrak{M}_1$, $\mathfrak{N}''_2 \simeq \mathfrak{M}_t$, and the natural representations of $\mathfrak{N}''_1$ and $\mathfrak{N}''_2$ have multiplicities 1 and (m - 1), respectively. In particular, $\dim X_{jj} \geq 1 + t^2$, $|X_{jj}| \geq 1 + (m - 1)t$. If $t > 1$, these inequalities are strict.

c) If $X_{ii} = S_m$, $|X_{jj}| = m$, $\dim X_{ij} = 2$, then $X_{jj} = S_m$.

Proof. We can assume that the rank of X is 2, i = 1, j = 2. In notations of 12, we have $\dim X_{12} = \sum_{p,q} \dim f'_p \mathfrak{A} f''_q$. Let us consider some algebra $\mathfrak{N}_i$. One has $\mathfrak{N}_i = f'_j \mathfrak{N}_i f'_j + f''_k \mathfrak{N}_i f''_k + f'_j \mathfrak{N}_i f''_k + f''_k \mathfrak{N}_i f'_j$, where $j = I'(i)$, $k = I''(i)$. The summands of this decomposition have dimensions r'^2_j, r''^2_k, $r'_j r''_k$, $r'_j r''_k$, respectively. We assume that some r'_j or r''_k can be equal to zero. Property (a) now takes the form

$$r'_j r''_k \leq (r'^2_j + r''^2_k)/2 ,$$

which is known to be true.

To prove (b) and (c), let us first note that in these cases $\mathfrak{A}' = \mathfrak{M}_1 \oplus \mathfrak{M}_1$, $\mu'_1 = 1$, $\mu'_2 = m - 1$, $r'_1 = r'_2 = 1$. There exists exactly one direct component of $\mathfrak{A}$, say $\mathfrak{N}_2$, which contains the second summand of $\mathfrak{U}$. By (12d), $\mu_2 = m - 1$. Remembering that $\mathfrak{N}_1$ is the subalgebra of $\mathfrak{A}$, defined in 6, and since

$$f_1' \mathfrak{N}_1 f_1' \simeq \mathfrak{M}_1, \; f_2' \mathfrak{N}_2 f_2' \simeq \mathfrak{M}_1 ,$$

we have $\dim f_1' \mathfrak{N}_1 f_1'' = 1$,

$$\mathfrak{A}_{12} = f_1' \mathfrak{N}_1 f_1'' \oplus f_2' \mathfrak{N}_2 f_2'' .$$

Hence $\dim f_2' \mathfrak{N}_2 f_2'' = t$, i.e.,

$$\mathfrak{N}_2'' = f_2'' \mathfrak{N}_2 f_2'' \simeq \mathfrak{M}_t .$$

In particular, $\dim X_{22} \geq 1 + t^2$. Since $\mu_2 = m - 1$, we have $\mu_2'' = m - 1$ (by 12d) and therefore

$$|X_{22}| \geq 1 + (m - 1)t .$$

If $t > 1$, then by 11 all inequalities are strict.

Let us now prove (c). In notations of (b), $t = 1$, i.e., by (b) $|X_{22}| \geq m$. Since $|X_{22}| = m$, one has by foregoing considerations $\mathfrak{A}'' \simeq \mathfrak{M}_1 \oplus \mathfrak{M}_1$, i.e., $\dim X_{22} = 2$, i.e., $X_{22} = S_m$.

14. Corollary. Let $X = (X_{ij})$ be a stationary graph. Suppose that the algebras $\mathfrak{A}(X_{ii})$ and $\mathfrak{A}(X_{jj})$ are commutative. Then

a) $\dim X_{ij} \leq \min(\dim X_{ii}, \dim X_{jj})$

b) If $\dim X_{ij} = \dim X_{ii}$, then $|X_{ii}| \leq X_{jj}$, $\dim X_{ii} \leq \dim X_{jj}$.

Proof. By the commutativity condition of $\mathfrak{A}' = \mathfrak{A}(X_{ii})$ and $\mathfrak{A}'' = \mathfrak{A}(X_{jj})$, we have in notations of 12: $r_i' = 1$ and $r_i'' = 1$. Hence by (12e), $r_i \leq 2$. If $r_i = 2$, then $\mathfrak{M}_2 \simeq \mathfrak{N}_i \supset \mathfrak{N}'_j \oplus \mathfrak{N}''_k$, where $j = I'(i)$, $k = I''(i)$. The number of $\mathfrak{N}_i$ for which $r_i = 2$ gives exactly the dimension of $\mathfrak{A}_{12}$ (since $\mathfrak{A}_{12} = f'\,\mathfrak{A}\,f''$, where f' and f'' are unities of $\mathfrak{A}'$, $\mathfrak{A}''$). This proves (a).

In case (b), we see that each $\mathfrak{N}'_j$ is contained in some $\mathfrak{N}_i \simeq \mathfrak{M}_2$. Hence by (a) and by (12c) (d), we have (b).

15. Let $\mathfrak{A}$ be a cellular algebra of rank 2 with matrix

$$X = \begin{pmatrix} X_{11} & X_{12} \\ X_{21} & X_{22} \end{pmatrix}$$

Let $\mathfrak{A} = \bigoplus_{i=1}^{t} \mathfrak{N}_i$, $\mathfrak{N}_i \simeq \mathfrak{M}_{r_i}$, and let μ_i be the multiplicity of the irreducible representation of $\mathfrak{N}_i$ in the natural representation of $\mathfrak{A}$. Finally let $V = \bigoplus_{i=1}^{t} \bigoplus_{j=1}^{\mu_i} V_{i,j}$ where V_{ij} are defined as in 9.

<u>Theorem</u> (compare [Fr 2]). Let $n = |X_{11}|$, $r = \dim X_{12}$, $e_1, \ldots, e_r \subset X_{12}$. For each $i \in [1, t]$ set correspondingly to 12:

$$p_i = r'_j \quad \text{if} \quad I'(i) = j$$

$$p_i = 0 \quad \text{if} \quad I'(i) = \emptyset$$

$$q_i = r''_k \quad \text{if} \quad I''(i) = k$$

$$q_i = 0 \quad \text{if} \quad I''(i) = \emptyset$$

Then

$$\frac{n^r \prod_{i=1}^{r} n_i}{\prod \mu^{p_i q_i}} = q \in \mathbf{Z}$$

Moreover, $q = a \cdot \bar{a}$, where a is an algebraic integer.

Proof of this theorem is analogous to the proof of Theorem 9. Let $X_{12} = \sum_{i=1}^{r} x_i e_i$, then

$$Sp(e_i \cdot e_{j'}) = \delta_{ij} \cdot n \cdot d(e_i), \quad i \in [1, r]$$

Further, we may assume that the matrices $M_{d,m,i} = U e_i U^{-1}$, $i \in [1, r]$ (see the proof of Theorem 9) are of the form

$$M_{d,m,i} = \begin{pmatrix} 0 & * \\ 0 & 0 \end{pmatrix} \begin{matrix} \} p_d \\ \} q_d \end{matrix}$$

Since $\Sigma p_d q_d = r$, the matrix $A = (a_i^\sigma)$, constructed as in 9, is a square matrix and we have the equality

$$N = \bar{A}' R A .$$

Taking the determinants we have

$$n^r \prod_{i=1}^{r} n_i = \left(\prod_{i=1}^{t} \mu_i^{p_i q_i}\right) a \cdot \bar{a}$$

where $a = \det A$.

It remains to show that a is an algebraic integer. Let us show this. Let $\mathfrak{N}_d$ be a direct summand of $\mathfrak{A}$; $\mathfrak{N}_d'$, $\mathfrak{N}_d''$ be the intersections of $\mathfrak{N}_d$ with $\mathfrak{A}'$ and $\mathfrak{A}''$ respectively.

One can assume that $\mathfrak{N}_d' \neq 0$, $\mathfrak{N}_d'' \neq 0$ since otherwise $p_d q_d = 0$. Let $V_{d,i}$ be a space of the irreducible representation of $\mathfrak{N}_d$. Then the matrices from $\mathfrak{N}_d'$ and $\mathfrak{N}_d''$ can be brought into the form

$$\begin{pmatrix} * & 0 \\ 0 & 0 \end{pmatrix}\begin{matrix} \}\ p_d \\ \}\ q_d \end{matrix} \quad \text{and} \quad \begin{pmatrix} 0 & 0 \\ 0 & * \end{pmatrix}\begin{matrix} \}\ p_d \\ \}\ q_d \end{matrix}$$

respectively. When $\mathfrak{N}'_d$ and $\mathfrak{N}''_d$ are of this form, it is evident from (12a) that matrices e'_i, $i = 1, 2, \ldots, r$, have the form shown on the preceding page. Let $V_{d,i} = V'_{d,i} \oplus V''_{d,i}$ be the corresponding decomposition of $V_{d,i}$ into the orthogonal direct sum. Let us choose a basis of $V'_{d,i}$ such that all matrices e_s, $e_s \subset X_{11}$ are written as matrices whose entries are algebraic integers (Lemma 8). Let the corresponding R-module be $\widetilde{V}'_{d,i}$.

Put $\widetilde{V}''_{d,i} = \sum_{t=1}^{r} R\, e_t\, \widetilde{V}'_{d,i}$. Then by (12a) $K\, \widetilde{V}''_{d,i} = V''_{d,i}$ and if $\{\xi_i\}$ is an R-basis of the R-module $\widetilde{V}'_{d,i} \oplus \widetilde{V}''_{d,i}$ then in this basis matrices e_i, $i \in [1, r]$, have algebraic integral entries. From this, our Theorem follows.

16. Corollary. Let $\mathfrak{A}$ be a cellular algebra of rank 2 with the matrix

$$X = \begin{pmatrix} X_{11} & X_{12} \\ X_{21} & X_{22} \end{pmatrix}$$

Let $N_i = |X_{ii}|$, $r_i = \dim X_{i1} + \dim X_{i2}$, and let $\{e_i\}$, $i = 1, \ldots, r_1 + r_2$, be the standard basis of $\mathfrak{A}$, $n_i = d(e_i)$. Then

$$\frac{N_1^{r_1 - 2} \cdot N_2^{r_2 - 2} \prod_{i=1}^{r_1 + r_2} n_i}{\prod_{i=1}^{t} \mu_i^{r_i^2}} = q \in \mathbb{Z}$$

Moreover, $q = a \cdot \bar{a}$, where a is an algebraic integer.

Proof. This is the product of the expressions for X_{11}, X_{22} given by Theorem 9 and for

X_{12}, X_{21} given by Theorem 15.

Remark. Conversely, the main part of Theorem 15 follows from this Corollary and Theorem 9. The only point which is not evident is that $q = a \cdot \bar{a}$ in Theorem 15.

17. Corollary. Let $X_{11} = S_n$, $\dim X_{12} = r$. Then

$$\left(\prod_{i=1}^{r} n_i \right) (n-1)^{-r+1} \in \mathbf{Z}$$

Proof. By (13b) $p_i \neq 0$ for only two values of i, say for $i = 1, 2$. Then one can assume that

$$\mu_1 = 1, \mu_2 = n - 1, p_1 = 1, p_2 = 1, q_1 = 1, q_2 = r - 1$$

Since $(n, n - 1) = 1$, our assertion follows from 15.

18. Remark. Let

$$X = \begin{pmatrix} X_{11} & X_{12} \\ X_{21} & X_{22} \end{pmatrix}$$

If $X_{11} = S_\mu$ and e_i is an element of a standard basis of X, $e_i \subset X_{12}$, then evidently $e_i \cdot e_i' = \lambda E + \mu I$, i.e., e_i is the matrix of some block-design. On the other hand, any symmetric block-design with matrix e can be considered as an element of a standard basis of some stationary graph X of rank 2, where, in addition, $X_{11} = X_{22} = S_m$.

This shows that cellular algebras can be considered as generalization of some popular combinatorial formations. It is plausible that the theory of block-designs could be developed in this direction. In [Hi 3] and [Bo 6], this approach is adopted.

Let us note in this connection that Theorem 10.2.2 from [Ha 3] coincides with our assertion (14b).

19. <u>Example</u>. Let $X_{11} = X_{22} = S_7$, $e_1 = E_7 \subset X_{11}$, $e_2 = \widetilde{I}_7 \subset X_{11}$, $e_3 = E_7 \subset X_{22}$, $e_4 = \widetilde{I}_7 \subset X_{22}$.

Let e_5 be the incidence matrix of the projective plane of order 7, $e_6 = I_7 - e_5$, $e_7 = e_5'$, $e_8 = e_6'$. Setting $X_{12} = x_5 e_5 + x_6 e_6$, $X_{21} = x_7 e_7 + x_8 e_8$, we see that

$$X = \begin{pmatrix} X_{11} & X_{12} \\ X_{21} & X_{22} \end{pmatrix}$$

is the matrix of a cellular algebra $\mathfrak{A}$ of rank 2. Evidently, $\mathfrak{A} = \mathfrak{N}_1 \oplus \mathfrak{N}_2$, $\mathfrak{N}_i \simeq \mathfrak{M}_2$, where the multiplicity of the irreducible representation of one of these summands (say of the first) is equal to unity, and that of the second is six. Let us write the elements a of $\mathfrak{A}$ in the form

$$a \longrightarrow \begin{pmatrix} a_{11} & a_{12} \\ a_{21} & a_{22} \end{pmatrix} \oplus \begin{pmatrix} b_{11} & b_{12} \\ b_{21} & b_{22} \end{pmatrix}$$

where the summands denote the projections of a onto $\mathfrak{N}_1$ and $\mathfrak{N}_2$, respectively, in an appropriate basis. (That is, the matrices $M_{d,m,i}$ have the above form.) We have

$$e_1 \longrightarrow \begin{pmatrix} 1 & 0 \\ 0 & 0 \end{pmatrix} \oplus \begin{pmatrix} 1 & 0 \\ 0 & 0 \end{pmatrix}, \quad e_2 \longrightarrow \begin{pmatrix} 6 & 0 \\ 0 & 0 \end{pmatrix} \oplus \begin{pmatrix} -1 & 0 \\ 0 & 0 \end{pmatrix}$$

$$e_3 \longrightarrow \begin{pmatrix} 0 & 0 \\ 0 & 1 \end{pmatrix} \oplus \begin{pmatrix} 0 & 0 \\ 0 & 1 \end{pmatrix}, \quad e_4 \longrightarrow \begin{pmatrix} 0 & 0 \\ 0 & 6 \end{pmatrix} \oplus \begin{pmatrix} 0 & 0 \\ 0 & -1 \end{pmatrix}$$

$$e_5 \to \begin{pmatrix} 0 & 3 \\ 0 & 0 \end{pmatrix} \oplus \begin{pmatrix} 0 & x \\ 0 & 0 \end{pmatrix}, \quad e_6 \to \begin{pmatrix} 0 & 4 \\ 0 & 0 \end{pmatrix} \oplus \begin{pmatrix} 0 & y \\ 0 & 0 \end{pmatrix}$$

$$e_7 \to \begin{pmatrix} 0 & 0 \\ 3 & 0 \end{pmatrix} \oplus \begin{pmatrix} 0 & 0 \\ \bar{x} & 0 \end{pmatrix}, \quad e_8 \to \begin{pmatrix} 0 & 0 \\ 4 & 0 \end{pmatrix} \oplus \begin{pmatrix} 0 & 0 \\ \bar{y} & 0 \end{pmatrix}$$

Let ζ be a primitive 7-th root of unity. Then we can assume

$$x = \zeta + \zeta^2 + \zeta^4$$

Since $e_5 + e_6 \in \mathfrak{N}_1$, $y = -x$. In addition $x \cdot \bar{x} = 2$.

Let us construct the matrix $Z = (a_i^\sigma)$ (cf. 9 and 15).

α, β, d \ i	1	2	3	4	5	6	7	8
1, 1, 1	1	6						
1, 1, 2	1	-1						
2, 2, 1			1	6				
2, 2, 2			1	-1				
1, 2, 1					3	4		
1, 2, 2					x	$-x$		
2, 1, 1							3	4
2, 1, 2							$\bar{x}$	$-\bar{x}$

We have $\det A = (-7) \cdot (-7) \cdot (-7x) \cdot (-7\bar{x}) = 7^4 \cdot x \cdot \bar{x} = 7^4 \cdot 2$, $\det A_1 = -7$, $\det A_2 = -7$ (cf. 15).

According to the proofs of Theorem 15, we have

$$\frac{\det A \cdot \det \bar{A}}{\det A_1 \cdot \det \bar{A}_1 \cdot \det A_2 \cdot \det \bar{A}_2} = \frac{7^8 \cdot 4}{7^2 \cdot 7^2} = 7^4 \cdot 4$$

Thus (compare 15),

$$\frac{n^r \cdot \prod n_i}{\prod \mu_i^{p_i q_i}} = \frac{7^2 \cdot 3 \cdot 4}{6} = 7^2 \cdot 2 \in \mathbf{Z}$$

20. Example. Let $\mathfrak{A}$ be a three-dimensional cell, $e_0 = E_n$, e_1, e_2 be its standard basis. Suppose we have $e_i' = e_i$. We can write

$$e_i^2 = a_i E_n + b_i e_i + c_i(\tilde{I}_n - e_i)$$

This shows that e_i, $i = 1,2$ are strongly regular graphs (cf. Section T and [Se 3]). Strongly regular graphs and, among other things, their spectral properties were intensively studied. The most striking result in this direction is contained in [Ca 2].

A geometric study of strongly regular graphs was strongly influenced by [Bo 2], [Bo 5]. Strongly regular graphs were also used to construct several sporadic groups [Hi 7], [Ti 1].

Below we shall consider three-dimensional cells from the point of view of this Section (cf. also [Hi 1]). In Section U, one can find examples of such cells.

Let us write

$$e_1 = n_1 f_0 + a \ f_1 + b \ f_2$$

$$e_2 = n_2 f_0 + a' \ f_1 + b' \ f_2$$

where $f_i^2 = f_i$, $f_i f_j = 0$ for $i \neq j$, are orthogonal idempotents of $\mathfrak{A}$, and $\mathrm{Sp}\, f_0 = 1$. Set

$$\mu_i = \mathrm{Sp}\, f_i$$

Then we have $\mu_1 + \mu_2 = n - 1$.

From $e_0 + e_1 + e_2 = I_n$ we conclude that $a' = -a - 1$, $b' = -b - 1$.

Note that $a \neq b$ since otherwise we would have $c_1 = n_1 f_0 + a(f_1 + f_2)$, whence $e_1^2 = n_1^2 f_0 + a^2(f_1 + f_0) = c_1 E_n + c_2 I_n$, a contradiction with the assumption that $\dim \mathfrak{A} = 3$. We have

$$\mathrm{Sp}\ e_1 = 0 = n_1 + a\mu_1 + b\mu_2$$

From this equality, and from $\mu_1 + \mu_2 = n - 1$, we deduce

$$a - b = -\frac{n_1 + b(n-1)}{\mu_1} = \frac{n_1 + a(n-1)}{\mu_2} \qquad (*)$$

We have next

$$e_1^2 = n_1^2 f_0 + a^2 f_1 + b^2 f_2$$

On the other hand

$$e_1^2 = n_1 e_0 + a_{11}^1 e_1 + a_{11}^2 e_2 = n_1(f_0 + f_1 + f_2)$$
$$+ a_{11}^1(n_1 f_0 + a f_1 + b f_2) + a_{11}^2(n_2 f_0 - (a+1)f_1 - (b+1)f_2) .$$

Combining the two preceding expressions for e_1^2, we get the following equation for a and b:

$$x^2 = n_1 + (a_{11}^1 - a_{11}^2)x - a_{11}^2$$

whence

$$x = \frac{(a_{11}^1 - a_{11}^2) \pm \sqrt{(a_{11}^1 - a_{11}^2)^2 + 4(n_1 - a_{11}^2)}}{2} \qquad (**)$$

This gives us (since $a \neq b$):

$$ab = - n_1 + a_{11}^2$$

$$a + b = a_{11}^1 - a_{11}^2$$

$$a - b = \pm \sqrt{(a_{11}^1 - a_{11}^2)^2 + 4(n_1 - a_{11}^2)}$$

Substituting this into the expression

$$(a - b)^2 = - \frac{n_1 + (a + b)n_1(n - 1) + ab(n - 1)^2}{\mu_1 \mu_2}$$

obtained by the multiplication of the two right-hand parts of (*), we get

$$(a - b)^2 = \frac{n\, n_1\, n_2}{\mu_1 \mu_2} \qquad (***)$$

which should be compared with the expression of Theorem 9.

Let us note that a, b are algebraic integers. If $a, b \in \mathbb{Z}$, we must have

$$(a_{11}^1 - a_{11}^2)^2 + 4(n_1 - a_{11}^2) = d^2, \; d \in \mathbb{Z}$$

Otherwise, let $a, b \notin \mathbb{Z}$. Let σ be the nontrivial automorphism of $\mathbb{Q}(a)$ over $\mathbb{Q}$. Since we have $e_1^\sigma = e_1$, we must have

$$\mu_1 = \mu_2, \; b = a^\sigma$$

Therefore, $\mu_1 = \mu_2 = \frac{n-1}{2}$. Therefore, $2 | (n - 1)$ and from (***), we conclude that

$$\frac{n_1 n_2}{\left(\frac{n-1}{2}\right)^2} \in \mathbf{Z}$$

But this implies that $n_1 = n_2 = \frac{n-1}{2}$, whence

$$(a - b)^2 = n .$$

Hence

$$a = m \pm \frac{\sqrt{n}}{2} , \quad b = m \mp \frac{\sqrt{n}}{2} .$$

This, together with the expression for $\mathrm{Sp}\, e_1$, gives us

$$m = \frac{1}{2}$$

that is,

$$a = \frac{1}{2} \pm \frac{\sqrt{n}}{2} , \quad b = \frac{1}{2} \mp \frac{\sqrt{n}}{2} .$$

Now apply the expression

$$ab = - n_1 + a_{11}^2$$

and get

$$- n_1 + a_{11}^2 = \frac{1 - n}{4}$$

which gives us

$$4 \mid n - 1, \quad a_{11}^{2} = \frac{n-1}{4}$$

Summarizing, we have the following

Assertion. If $\mathfrak{A}$ is a three-dimensional cell of degree n with the standard basis

$$e_0 = E_n, \; e_1, \; e_2$$

and if $e_i' = e_i$, then either $n = 4q + 1$, $n_1 = n_2 = 2q$, $a_{11}^{2} = q$, or

i) $(a_{11}^{1} - a_{11}^{2})^2 + 4(n_1 - a_{11}^{2}) = d^2$, $d \in \mathbf{Z}$,

ii) The eigenvalues of a_1 are $a_i = \dfrac{(a_{11}^{1} - a_{11}^{2}) \pm d}{2}$, $i = 1, 2$, and n_1 with the multiplicities $\left|\dfrac{n_1 + a_i(n-1)}{d}\right|$ and 1 respectively.

M. SOME MODIFICATIONS OF STABILIZATION.

1. This Section is the first where we are concerned with algorithmic questions. It can be considered as a setting of a stage for the treatment of such questions. The procedure of stabilization described in Section C is insufficient for a description of algorithms (but more convenient for aims of Sections D-L). We describe here a modification which makes use of the order of the elements of the adjacency matrix of a graph. Some additional modifications are also given. The methods described below are used in Sections N, O, R.

2. <u>Correspondence: geometrical graph-matrix whose entries are independent variables</u>

2.1. <u>Definition</u>. Let $A = (a_{ij})$ be a $(n \times n)$-matrix whose elements belong to a partially ordered set M. The order in M is denoted by $\geq$, $\leq$. If, for $a, b \in M$, the order is not defined, we write $a \geq b$, $b \geq a$. If $a \neq b$, $a \geq b$, $b \not\geq a$, we write $a > b$. We assume that the partial order on M satisfies the following condition:

If $a > b$, $b \geq c$, $c \geq b$, then $a > c$. (This is, in particular, a justification of our notation $a \geq b$, $b \geq a$ for a pair with undefined order.)

Since this condition is preserved throughout all our actions on graphs, we assume henceforth without mentioning that all partially ordered sets satisfy this condition.

Let $X(A) = (x_{ij})$ be a graph (in the sense of C1) defined by

a) $x_{ii} = x_{kk}$ if and only if $a_{ii} = a_{kk}$;

b) $x_{ii} \geq x_{kk}$ if and only if $a_{ii} \geq a_{kk}$;

c) $x_{ii} > x_{kd}$ for all i and all $k \neq d$;

d) $x_{ij} = x_{kd}$, $i \neq j$, $k \neq d$, if and only if $a_{ij} = a_{kd}$;

e) $x_{ij} \geq x_{kd}$, $i \neq j$, $k \neq d$, if and only if $a_{ij} \geq a_{kd}$;

f) The variables entering in $X(A)$ are numbered from 1 to $\dim X(A)$, and this numeration agrees as far as possible with the partial order of the variables, that is, if $x_{ij} = x_t$, $x_{kd} = x_s$, $x_t > x_s$, then $t > s$.

2.2. If $A_1, A_2, \ldots, A_m$ are matrices with entries from M, we set $X(A_t) = (x^t_{ij})$ and define

$$X(A_1 \cap \ldots \cap A_m) = (x_{ij})$$

in the following manner

a) $x_{ij} = x_{kd}$ if and only if $x^t_{ij} = x^t_{kd}$ for all t;

b) $x_{ij} \geq x_{kd}$ if and only if $(x^1_{ij}, \ldots, x^m_{ij}) \geq (x^1_{kd}, \ldots, x^m_{kd})$;

c) The same as in (2.1f).

This construction is used for instance in N 3.3 and in O 4.9, 4.11.

2.3. If M is a linearly ordered set, e.g. $\mathbb{R}$, $\mathbb{Z}$, then the variables of the matrix X(A) are linearly ordered.

If A is the adjacency matrix of a simple geometrical graph, then $M = \{0, 1\}$ and M is a linearly ordered set $(1 > 0)$. Thus, the variables of X(A) are linearly ordered in this case.

Such an approach can turn helpful when one uses cellular algebras not only for the study of graphs, but also for the study of orbits of Sym(n) on $V \otimes V^*$, where V is a module over a ring (cf. AE 1.2).

3. Stabilization

3.1. Let X be a graph whose variables x_i are partially ordered. Let us define a partial order of monomials of degree 2 on x_i in the following manner. (Recall that independent variables do not commute, cf., C 1.) Set $x_i x_j \geq x_s x_t$ if and only if $x_i \geq x_s$ or $x_i \leq x_s$, $x_j \geq x_t$.

Let us extend this partial order lexicographically to all homogeneous polynomials of degree 2. Let us further choose two additional variables λ and μ, and assume that $\lambda > \mu$, and that both λ and μ are strictly greater than all variables x_i.

3.2. Define the graph $X \circ X = (y_{ij})$ as follows (compare C 4):

a) $y_{ij} = y_{kd}$ if and only if $\Sigma_r x_{ir} x_{rj} + \lambda x_{ij} + \mu x_{ji} = \Sigma_r x_{kr} x_{rd} + \lambda x_{kd} + \mu x_{dk}$;

b) $y_{ij} \geq y_{kd}$ if and only if $\Sigma_r x_{ir} x_{rj} + \lambda x_{ij} + \mu x_{ji} \geq \Sigma_r x_{kr} x_{rd} + \lambda x_{kd} + \mu x_{dk}$;

c) the variables of $X \circ X$ are numbered from 1 to dim X, and this numeration agrees with the order of the variables of $X \circ X$ (in the sense of 2.1f).

Set $X^0 = X$, $X^{i+1} = X^i \circ X^i$. If $\dim X^{i-1} < \dim X^i = \dim X^{i+1}$, we call X^i the <u>stabilization</u> of X and write Stab $X = X^i$.

3.3. <u>Remarks</u>. If the variables of X are linearly ordered, then the variables of Stab X are also linearly ordered. The need to use this operation also for partially ordered variables arises, for example, in the study of the kernel (cf. next Section).

3.4. <u>Lemma</u>. One has $\mathrm{Stab}(\sigma X \sigma^{-1}) = \sigma(\mathrm{Stab}\ X)\sigma^{-1}$ for $\sigma \in \mathrm{Sym}\ V(X)$.

<u>Proof.</u> Evident.

4. <u>Simultaneous stabilization</u>

4.1. It is possible that the stabilizations Stab X and Stab Y of two graphs X and Y contain the same variables which have the same order. For instance, if the variables of X and Y are linearly ordered, our assumption implies only that dim Stab X = dim Stab Y. In this case, the coincidence of variables of Stab X and Stab Y does not imply that those variables have the same origin. In some cases, however, it is convenient to secure that the "history" (or "genealogy") of equally named variables would be the same. The corresponding definitions are given below.

4.2. Let $\{A_1, \ldots, A_m\}$ be an ordered set of $(n \times n)$-matrices whose entries lie in a partially ordered set M, $A_s = (a^s_{ij})$. We shall denote by $\dim\{A_1, \ldots, A_m\}$ the number of different entries of these matrices. Define the set of graphs $X(A_1,\ldots,A_m) = \{X_1,\ldots,X_m\}$ where $X_s = (x^s_{ij})$ by

a) $x^s_{ii} = x^t_{kk}$ if and only if $a^s_{ii} = a^t_{kk}$, $s, t \in [1, m]$, $i, k \in [1, n]$;

b) $x^s_{ii} \geq x^t_{kk}$ if and only if $a^s_{ii} \geq a^t_{kk}$, $s, t \in [1, m]$, $i, k \in [1, n]$;

c) $x^s_{ii} > x^t_{kd}$ for all $s, t \in [1, m]$, all $i, k, d \in [1, m]$, $k \neq d$;

d) $x^s_{ij} = x^t_{kd}$, $i \neq j$, $k \neq d$, if and only if $a^s_{ij} = a^t_{kd}$, $s, t \in [1, m]$;

e) $x^s_{ij} \geq x^t_{kd}$, $i \neq j$, $k \neq d$, if and only if $a^s_{ij} \geq a^t_{kd}$, $s, t \in [1, m]$;

f) the variables of $X(A_1, \ldots, A_m)$ are numbered from 1 to dim $X(A_1, \ldots, A_m)$, and this numeration agrees with the order of the variables (cf. 2.1f).

4.3. Let $\{X_1, \ldots, X_m\}$ be an ordered set of graphs of the same degree n, whose variables are partially ordered.

Set $\{X_1, \ldots, X_m\} = \{X_i\}^0$ and if $\{X_i\}^q = \{Y_i\}$, $Y_t = (x^t_{ij})$, then define $\{X_i\}^{q+1} = \{Z_i\}$, $Z_t = (z^t_{ij})$, by

a) $z^t_{ij} = z^s_{kd}$ if and only if $\Sigma_r\, x^t_{ir}\, x^t_{rj} + \lambda\, x^t_{ij} + \mu\, x^t_{ji} = \Sigma_r\, X^s_{kr}\, x^s_r + \lambda\, x^s_{kd} + \mu\, x^s_{dk}$;

b) $z^t_{ij} \geq z^s_k$ if and only if $\Sigma_r\, x^t_{ir}\, x^t_{rj} + \lambda\, x^t_{ij} + \mu\, x^t_{ji} \geq \Sigma_r\, x^s_{kr}\, x^s_{rd} + \lambda\, x^s_{kd} + \mu\, x^s_{dk}$;

c) the variables of $\{X_i\}^{q+1}$ are numbered from 1 to dim $\{X_i\}^{q+1}$, etc. (see 2.1f).

If $\dim\{X_i\}^{q-1} < \dim\{X_i\}^{q+1} = \dim\{X_i\}^q$, we say that $\{X_i\}^q$ is the <u>simultaneous stabilization</u> of X_i and write $\mathrm{Stab}\{X_i\} = \{X_i\}^q$.

4.4. <u>Proposition</u>. Let $\{X_i\}$ be a set of graphs of the same degree and $\mathrm{Stab}\{X_i\} = \{X_i\}$. If X_k and X_d have the same composition, $X_k = \Sigma_{i \in I}\, x_i\, e^k_i$, $X_d = \Sigma_{i \in I}\, x_i\, e^d_i$, then the identity map of I into itself is a weak equivalency of X_k and X_d.

<u>Proof</u>. Let $e^t_i \cdot e^t_j = \Sigma_s\, a^{ts}_{ij}\, e^t_s$. If $a^{ks}_{ij} \neq a^{ds}_{ij}$ for some triple (s, i, j) then by the definition of the simultaneous stabilization, X_k and X_d would have different composition (cf. 4.3a).

4.5. **Lemma.** Let $\{X_i\}$ be a set of graphs of the same degree n. One has $\mathrm{Stab}\{\sigma_i X_i \sigma_i^{-1}\} = \{\sigma_i(\mathrm{Stab}\, X_i)\sigma_i^{-1}\}$ for $\sigma_1, \sigma_2, \ldots \in \mathrm{Sym}\, n$.

Proof. Evident.

N. KERNELS AND STABILITY WITH RESPECT TO KERNELS.

The constructions of this Section are motivated by permutation group theory. Explicitly, consider a permutation group G acting on the set X. Let Y be one of the orbits of G. Let G_Y be the pointwise stabilizer of the points of Y in G. In this situation, the construction of this Section aims at the description of $\mathcal{J}(G_Y, X)$ in terms of $\mathcal{J}(G, X)$ (cf. Section F). This shows the importance of taking kernels.

We do not use this operation in the algorithm of Section R. However, it can be used, at least at heuristical level.

1. Let X be a stationary graph. It is convenient to assume in this Section that the variables of X are linearly ordered. For instance, this order can be chosen arbitrary. Let us, however, not that the algorithm (cf. M 2) which constructs for a geometrical graph Γ the corresponding stationary graph Stab $X(\Gamma)$, leads just to a stationary graph with linearly ordered variables.

2. <u>Definition of the kernel of X on W</u>

Let $X = (X_{ij}) = (x_{mn}) = \Sigma\, x_i\, e_i$ be a stationary graph, and let the x_i be linearly ordered. Let $W = \bigcup_{i \in J} V(X_{ii})$.

Let $\overline{X} = (\overline{x}_{pq})$ be the graph obtained from X by substituting for all x_{qq}, $q \in W$, the new variables y_i where

a) $\overline{x}_{pp} \neq \overline{x}_{qq}$ for all $p \neq q \in W$;

b) $\overline{x}_{qr} > \overline{x}_{ts}$ if and only if $x_{qr} > x_{ts}$, $t, s, q, r \in V(X)$.

Thus the variables of $\overline{X}$ are partially ordered. Let us note (and this is important) that for the variables $\overline{x}_{pp}$ and $\overline{x}_{qq}$, $p \neq q \in V(X_{ii}) \subset W$, the order is not defined.

Set

$$\varkappa_W(X) = \text{Stab}\ \overline{X}$$

This matrix is called the <u>kernel</u> of X on W.

2.1. Proposition. Aut $\varkappa_W(X) = \{g \in \text{Aut } X : g|W = 1\}$ (whence the name "kernel").

Proof. The right-hand side evidently contains the left-hand side. The opposite inclusion follows by C 8.2 and by the evident equality: Aut $\overline{X} = \{g \in \text{Aut } X : g|W = 1\}$.

2.2. Remark. $\varkappa_W(X)$ is equivalent to the stabilization of the intersection (in the sense of M 2.2) of the graphs $\lambda_i(X)$, $i \in W$, cf. next Section.

2.3. Remark. Geometrically the construction of the kernel with respect to W means that we assign to all vertices from W pairwise different colors (and different from the colors already used in X). Clearly in this approach we cannot set any invariant order on the vertices of W. So we are forced to assume that the new colors are not ordered. After this repainting, we stabilize the new graph. The next undertaking (cf. Subsection 3, below) consists in finding whether some of the vertices or edges which were indistinguishable in the original graph behave differently in the kernel. If they behave differently, we can invariantly introduce new colors in X itself.

3. Definition of the stabilization with respect to kernel

3.1. Let us write $X_{ij} > X_{kd}$ if the greatest variable of X_{ij} is greater than the greatest variable of X_{kd}. Let $\varkappa_W(X) = Y = (Y_{pq})_{p,q\in I} = \Sigma\, y_i\, f_i$.

Let $I_i = \{p : Y_{pp} \subset X_{ii}\}$ and denote by Π the partition $I = \bigcup I_i$. Let $\mu_{i,\Pi}(Y_{p,q})$ be the vectors defined in E 6. Set $\nu_{pq} = (\mu_{2,\Pi}(Y_{pp}), \mu_{2,\Pi}(Y_{qq}), \mu_1(Y_{pq}))$.

3.2. Let us define a partial order of the blocks Y_{pq}. The conditions of the ordering are written down in the order of priority.

a) $Y_{pq} \subset X_{ij}$, $Y_{st} \subset X_{kd}$, if $X_{ij} > X_{kd}$ then $Y_{pq} > Y_{st}$;

b) if $\nu_{pq} > \nu_{st}$ then $Y_{pq} > Y_{st}$;

c) if $Y_{pp} > Y_{qq}$ then $Y_{pr} > Y_{qs}$ for all r, s;

d) if $Y_{pp} > Y_{qq}$ then $Y_{rp} > Y_{sq}$ for all r, s.

If these conditions did not determine an order among Y_{pq} and Y_{rs}, we shall write $Y_{pq} \approx Y_{rs}$.

3.3. Let us define a partial order of f_i's. The conditions are written in the order of priority.

a) $f_i \subset e_r$, $f_j \subset e_t$, if $x_r > x_t$ then $f_i > f_j$;

b) $f_i \subset Y_{pq}$, $f_j \subset Y_{st}$ if $Y_{pq} > Y_{st}$ then $f_i > f_j$;

c) let $X_i = \text{Stab}\, X(f_i \cap X)$ (cf. L 2.2) if $X_i > X_j$ then $f_i > f_j$.

If these conditions do not determine an order among f_i and f_j, we shall write $f_i \approxeq f_j$.

Remark. Clearly (c) is stronger than (a). (a) is included here to make references more convenient.

3.4. Let $\tilde{f}_i = \Sigma_{f_i \approxeq f_j} f_j$. Let K be a set of indices such that $\tilde{f}_i \neq \tilde{f}_j$ for any $i \neq j \in K$, and such that for every $\tilde{f}_i$ there exists $j \in K$ such that $\tilde{f}_i = \tilde{f}_j$. It is evident that $\tilde{f}_i \cap \tilde{f}_j \neq 0$ implies $\tilde{f}_i = \tilde{f}_j$. Hence $\tilde{f}_i \cap \tilde{f}_j = 0$ for all $i \neq j \in K$.

Set $\tilde{X} = \Sigma_{i \in K} \tilde{x}_i \tilde{f}_i$ and $\tilde{x}_i > \tilde{x}_j$ if $f_i > f_j$, $i, j \in K$. Put

$$\rho_W(X) = \text{Stab}\, \tilde{X}$$

We shall call this matrix the <u>stabilization of X with respect to the kernel on W</u>.

By the remarks in the beginning of this section, $\rho_W(X)$ is the matrix with linearly ordered variables.

3.5. <u>Proposition</u>. The graph $\rho_W(X)$ is defined invariantly, i.e., if $g \in \text{Sym}\, V(X)$ then $\rho_{gW}(g X g^{-1}) = g(\rho_W(X))g^{-1}$.

<u>Proof.</u> Evident.

3.6. <u>Proposition</u>. $\text{Aut}\, X = \text{Aut}\, \rho_W(X)$.

<u>Proof.</u> Evident.

4. <u>Definition and properties of stable graphs</u>

4.1. We say that X is <u>stable with respect to the kernel on</u> W if

$\dim X = \dim \rho_W(X)$, (that is, if X and $\rho_W(X)$ are equivalent). If $|X_{ii}| = m = rt$ and if for all $Y_{pp} \subset X_{ii}$ one has $|Y_{pp}| = r$, then we say that $\varkappa_W$ <u>decomposes</u> X_{ii} into t (equal) parts of degrees r. If $t = m$, we say that $\varkappa_W$ <u>splits</u> X_{ii}, and if $t = 1$, we say that $\varkappa_W$ <u>does not decompose</u> X_{ii}.

4.2. <u>Proposition</u>. Let a stationary graph X be stable with respect to the kernel on W, $Y = \varkappa_W(X) = (Y_{pq}) = \Sigma\, y_m f_m$; $X = (X_{ij}) = \Sigma\, x_t e_t$.

a) If $Y_{pp}, Y_{qq} \subset X_{ii}$, then $Y_{pp} \approxeq Y_{qq}$. In particular, $|Y_{pp}| = |Y_{qq}|$, $\dim Y_{pp} = \dim Y_{qq}$, $\mu_{2,\Pi}(Y_{pp}) = \mu_{2,\Pi}(Y_{qq})$;

b) $\{V(Y_{pp})\}_{Y_{pp} \subset X_{ii}}$ is an imprimitivity system for X_{ii};

c) If $f_s, f_t \subset e_r$ then $f_s \approxeq f_t$ and $e_r = \tilde{f}_s$.

In particular $d(f_s) = d(f_t)$, and natural weak equivalency of $\mathfrak{a}(f_s \cap X)$ and $\mathfrak{a}(f_t \cap X)$ is well-defined.

<u>Proof</u>. (a) and (c) follow directly from the condition of the stability of X and from 3.2, 3.3.

Let us prove (b). Let $Y_{pp} \subset X_{ii}$, $f_i \subset Y_{pp}$. By 3.2a,b, if $f_j \approxeq f_i$ then $f_j \subset Y_{qq}$, $Y_{qq} \approxeq Y_{pp}$. Hence, $\tilde{f}_i \subset \operatorname{diag}\,(Y_{pp})_{Y_{pp} \subset X_{ii}}$. In particular, $\tilde{f}_i$ is disconnected. The set of vertices of the connected components of $\Sigma_{f_i \subset Y_{pp}} \tilde{f}_i$ coincides with sets $V(Y_{pp})$, whence (b).

4.3. <u>Corollary</u>. Let X be as in 4.2. If the cell X_{ii} is primitive, then either $\varkappa_W$ splits X_{ii} or $\varkappa_W$ does not decompose X_{ii}.

<u>Proof</u>. Since in a primitive cell all normal subcells are trivial ones, it follows from 4.2b that $|Y_{pp}| = 1$ or $|Y_{pp}| = |X_{ii}|$ for any $Y_{pp} \subset X_{ii}$. Q.E.D.

4.4. <u>Lemma</u>. Let X be as in 4.2, $X = (X_{ij})_{i,j \in I}$. Let $I = I_1 \cup I_2$ where $\varkappa_W$ splits X_{ii}, $i \in I_1$, and $\varkappa_W$ does not decompose X_{ii}, $i \in I_2$. Then

$$X = X_1 \oplus X_2 \quad \text{where} \quad X_t = (X_{ij})_{i,j \in I_t}$$

Proof. We must show that $X_{ij} = \text{const}$, $i \in I_1$, $j \in I_2$. We may assume that $W = \bigcup_{i\in I_1} V(X_{ii})$, $W = [1, r]$. Set $Y = (Y_{pq})$. Clearly, Y_{pq} is a row of length $|Y_{qq}|$ for any $p \in [1, r]$. Let $q > r$, $V(X_{ii}) = V(Y_{qq})$. If $p \in V(X_{jj})$, $j \in I_1$, then it follows by the previous remarks that X_{ji} contains a constant row, namely Y_{pq}. This means that $X_{ji} = \text{const}$. Q.E.D.

4.5. Corollary. Let X be as in 4.2, $X = (X_{ij})_{i,j\in I}$. Suppose that cells X_{ii}, $i \in I_1$, are primitive and set $J = I - I_1$, $W = \bigcup_{i\in J} V(X_{ii})$. Then either $\varkappa_W$ splits X or $X = X_1 \oplus X_2$ for appropriate X_1 and X_2.

Proof is obtained by successive application of 4.3 and 4.4.

5. Variants

5.1. Many of the constructions given in this Section can be strengthened. Such constructions were not introduced above because we know of no assertion which uses their full power. Actually, Proposition 4.2 (which also does not use all given constructions) is completely sufficient for our purposes.

5.2. $\widetilde{\varkappa}_W(X)$. It is possible to consider instead of $\varkappa_W(X)$ the matrix $\widetilde{\varkappa}_W(X)$ which is defined in the following manner:

Let $\widetilde{W} = \{i \in V(\varkappa_W(X)) : \exists\, p,\ V(Y_{pp}) = i\}$. In the matrix X, replace the submatrix $(x_{ij})_{i,j\in V(X)-\widetilde{W}}$ by the corresponding submatrix of $\varkappa_W(X)$. Call the obtained matrix $\widetilde{X}$, and put $\widetilde{\varkappa}_W(X) = \text{Stab}\, \widetilde{X}$. The constructions of Subsection 3 are easily carried over to this case. In general, $\widetilde{\varkappa}_W(X)$ gives more information than $\varkappa_W(X)$ since the restriction of $\widetilde{\varkappa}_W(X)$ on W may be non-split.

5.3. Strengthening of 3.2 and 3.3. The methods can be strengthened by repetition up to stabilization. It is possible, moreover, to consider stronger invariants (cf., e.g., E 6.3) in place of $\mu_{2,\Pi}(Y_{pq})$ and $\mu_1(Y_{pq})$.

5.4. Strengthening of $\rho_W(X)$. Instead of stabilization with respect to $\varkappa_W(X)$, it is possible to stabilize with respect to all matrices which arise from X (cf. 2) in the process of its stabilization cf. M 2). This method may lead to stronger conditions.

6. Examples

6.1. I do not have examples where $\rho_W(X) \neq X$.

6.2. Consider the graph

$$X = \begin{pmatrix} x & y & z & u & u & v & v & w & w \\ z & x & y & v & v & w & w & u & u \\ y & z & x & w & w & u & u & v & v \\ a & b & c & p & q & r & s & m & n \\ a & b & c & q & p & s & r & n & m \\ b & c & a & m & n & p & q & r & s \\ b & c & a & n & m & q & p & s & r \\ c & a & b & r & s & m & n & p & q \\ c & a & b & s & r & n & m & q & p \end{pmatrix}$$

Take $W = \{1, 2, 3\}$. Then

$$\varkappa_W(X) = \begin{pmatrix} x_1 & x_2 & x_3 & y_1 & y_1 & y_4 & y_4 & y_7 & y_7 \\ x_4 & x_5 & x_6 & y_2 & y_2 & y_5 & y_5 & y_8 & y_8 \\ x_7 & x_8 & x_9 & y_3 & y_3 & y_6 & y_6 & y_9 & y_9 \\ z_1 & z_2 & z_3 & a_1 & b_1 & u_1 & v_1 & u_2 & v_2 \\ z_1 & z_2 & z_3 & b_1 & a_1 & v_1 & u_1 & v_2 & u_2 \\ z_4 & z_5 & z_6 & u_3 & v_3 & a_2 & b_2 & u_4 & v_4 \\ z_4 & z_5 & z_6 & v_3 & u_3 & b_2 & a_2 & v_4 & u_4 \\ z_7 & z_8 & z_9 & u_5 & v_5 & u_6 & v_6 & a_3 & b_3 \\ z_7 & z_8 & z_9 & v_5 & u_5 & v_6 & u_6 & b_3 & a_3 \end{pmatrix}$$

There is no order relation inside the following groups of the variables (x_1, x_5, x_9), (x_2, x_6, x_7), (x_3, x_4, x_8) (y_1, y_6, y_8) (y_2, y_4, y_9), (y_3, y_5, y_7), (v_1, v_4, v_5), (v_2, v_3, v_6), (u_1, u_4, u_5), (u_2, u_3, u_6), (a_1, a_2, a_3), (b_1, b_2, b_3), (z_1, z_5, z_9), (z_2, z_6, z_7), (z_3, z_4, z_8).

We have

$$\rho_W(X) = X$$

6.3. In this example we use the constructions of G 4 and J 4.3. Let $Y_1, \ldots, Y_n$ be naturally weakly isomorphic cells of degree m, and let Z be a cell of degree n. Set

$$X_{11} = (Y_1, \ldots, Y_n) \text{wr } Z$$

Then by G 4.4, X_{11} has a normal subcell $\mathcal{L}$ such that $X_{11}/\mathcal{L} \sim Z$. Using J 4.3 we can construct

$$X = \begin{pmatrix} X_{00} & X_{01} \\ X_{10} & X_{11} \end{pmatrix}$$

where $X_{00} \sim Z$ and $X_{01} \sim x\, E_n \otimes I_{m,1} + y\, \tilde{I}_n \otimes I_{m,1}$. Set $W = V(X_{00})$. Then $\varkappa_W(X) = \varkappa_W(X_{00}) \oplus \tilde{\bigoplus}_{i=1}^n \tilde{Y}_i$, where the $\tilde{Y}_i$'s have disjoint composition but $\tilde{Y}_i \sim Y_i$. Of course, $\varkappa_W(X_{00})$ is split. We have

$$\rho_W(X) = X$$

O. DEEP STABILIZATION.

1. Examples (cf., e.g., Section U) show that Stab X is a good, but insufficient invariant of X. To make this invariant more powerful we apply deep stabilization. There are several ways to introduce deep stabilization. We discuss here in more-or-less detail one approach (others are briefly discussed at the end of this Section, cf. also Section AD).

The construction described below is modeled on permutation groups. Let G be a group of permutations of a set V, and x a point of V. How can one describe $\mathfrak{Z}(G_x, V)$ in terms of $\mathfrak{Z}(G, V)$? Our graphs $\lambda_x(X)$ are analogues of $X(\mathfrak{Z}(G_x, V))$ in the case $X = X(\mathfrak{Z}(G, V))$.

The construction of this Section is used in the description of the algorithm of Section R. This latter algorithm uses stabilization not only with respect to $\{\lambda_m(X)\}$, but also with respect to more refined daughter systems (e.g., with respect to $\{\lambda_m(X)\}$ stabilized up to depth k). This forces us to consider general daughter systems (cf. 3.1 below).

2. Invariant algorithms

2.1. Let $\mathfrak{X}$ be the set of graphs and $\mathcal{A}$ be an algorithm on graphs, i.e., a (computable) function from $\mathfrak{X}$ into $\mathfrak{X}$.

An algorithm $\mathcal{A}$ is called invariant if for every substitution σ of the set of vertices of $X \in \mathfrak{X}$ one has

$$\mathcal{A}(\sigma X \sigma^{-1}) = \sigma(\mathcal{A}(X))\sigma^{-1}.$$

$\mathcal{A}$ is called correct on X if $\sigma X \sigma^{-1} = \tau X \tau^{-1}$ implies $\mathcal{A}(\sigma X \sigma^{-1}) = \mathcal{A}(\tau X \tau^{-1})$.

Lemma. An algorithm $\mathcal{A}$ is correct on X if and only if it is invariant with respect to all $\sigma \in \mathrm{Aut}\, X$, i.e., if

$$\mathcal{A}(\sigma X \sigma^{-1}) = \mathcal{A}(X) \quad \text{for } \sigma \in \mathrm{Aut}\, X.$$

<u>Proof.</u> Evident.

2.2. The algorithm $X \longrightarrow \text{Stab}\, X$ is invariant, cf. C 8.2, M 3.4.

2.3. It is possible also to define invariant algorithms from $\mathfrak{X}$ into $\mathfrak{X} \times \mathfrak{X} \times \ldots \times \mathfrak{X}$. Below, such an algorithm is considered, and it is shown how to use it to construct an invariant algorithm from $\mathfrak{X}$ into $\mathfrak{X}$, which is stronger than Stab.

3. <u>Daughter systems. Systems</u> $\{\lambda_i(X)\}$

3.1. Let $X = (X_{ij})$ be a stationary graph, $W = V(X_{tt})$. A system of stationary graphs $\{X_i\}_{i\in W}$ is called a <u>daughter system</u> of X with respect to W. We write $\{X_i\} = D_W(X)$.

Let $\mathcal{A}$ be an algorithm which constructs for the pair consisting of the stationary graph $X = (X_{ij})$ and the set $W = V(X_{tt})$, a daughter system $D_W(X) = \{X_i\}_{i\in W}$.

$\mathcal{A}$ is called <u>invariant</u> if $\mathcal{A}(\sigma X \sigma^{-1}) = \{\sigma X_{\sigma^{-1}(i)} \sigma^{-1}\}_{\sigma(i)\in\sigma(W)}$ for all $\sigma \in \text{Sym}\, V(X)$.

The corresponding daughter system $D_W(X)$ is then said to be <u>defined invariantly</u>.

3.2. The principal example of a daughter system which is used below is the system $\{\lambda_m(X)\}_{m\in W}$.

Let $X = (x_{ij})$ and take $m \in W$. Let $\overline{X}_m = (\overline{x}_{mij})$ be the graph defined in the following manner. Let y be a new variable, $y > x_{ij}$ for all i,j. Set $\overline{x}_{mij} = x_{ij}$ if $i \neq m$ or $j \neq m$, $\overline{x}_{mmm} = y$.

Now set (cf. M 4.2, 4.3) $\{\lambda_m(X)\} = \text{Stab}\{\overline{X}_m\}_{m\in W}$ (simultaneous stabilization).

We say that $\lambda_m(X)$ is obtained from X by <u>deleting</u> the m-th row (column).

If $\mathcal{A}$ is an invariant algorithm on graphs, then $\{\mathcal{A}(\lambda_m(X))\}_{m\in W}$ also is an example of an invariantly defined daughter system.

3.2.1 Let us note that if the variables of X are linearly ordered, then the variables of all graphs $\lambda_i(X)$ are linearly ordered.

3.2.2. Geometrically, $\lambda_i(X)$ is the graph obtained from X in the following way.

Choose a new color (which is not used in X), paint the i-th vertex of X in this color, and then stabilize.

3.3. Lemma. Aut $\lambda_m(X) = (\text{Aut } X)_m$ (= the stationary group in Aut X of the point m).

Proof. The inclusion $(\text{Aut } X)_m \supseteq \text{Aut } \lambda_m(X)$ follows from the definition of $\lambda_m(X)$. The reverse inclusion follows by C 10 from the obvious equality

$$(\text{Aut } X)_m = \text{Aut } \overline{X}_m$$

where $\overline{X}_m$ is as in 3.2.

3.4 Theorem. a) Let $\tau \in \text{Sym } V(X)$ be an isomorphism of $\lambda_t(X)$ on $\lambda_s(X)$. Then $\tau \in \text{Aut } X$ and $\tau t = s$.

b) If $\tau \in \text{Aut } X$, $t, s \in W$, $\tau t = s$, then τ is an isomorphism of $\lambda_t(X)$ on $\lambda_s(X)$.

Proof. (b) is evident. Let us prove (a). We must show that $\tau^{-1} X \tau = X$. Let $X = (x_{ij})$, $\lambda_t(X) = (y_{ij})$, $\lambda_s(X) = (z_{ij})$. Note that by the properties of the simultaneous stabilization, the equality $y_{ij} = z_{kl}$ implies equality $x_{ij} = x_{kl}$. Since $\tau^{-1} \lambda_t(X) \tau = \lambda_s(X)$, we have $z_{\tau i, \tau j} = y_{ij}$ for all i,j. By the above remark, it follows that $x_{\tau i, \tau j} = x_{ij}$. Q.E.D.

3.5. Theorem. Suppose that a partition $W = \bigcup W_i$ is such that $p, q \in W_i$ if and only if $\lambda_p(X)$ and $\lambda_q(X)$ have equal composition. If $\lambda_p(X) = R$ for all $p \in W_m$, then W_m is an orbit of the group Aut X and Aut X acts on W_m faithfully and fixed-point-free.

Proof. By 3.4b, Aut X preserves W_m. By 3.4a and E 5.6, Aut X is transitive on W_m. The last assertion follows from 3.3 by the condition $\lambda_p(X) = R$ for all $p \in W_m$.

3.6. Remark. In the case when $X = \mathfrak{Z}(G, V)$, G a permutation group of V, I do not know whether $\lambda_i(X) = X(\mathcal{J}(G_i, V))$ (although one has 3.3).

4. Stabilization with respect to the system $\{\lambda_m(X)\}_{m\in W}$

4.1. Define $\overline{X} = (\overline{x}_{ij})$ in the following manner:

a) $\overline{x}_{ij} = x_{ij}$ for $i \neq j$;

b) $\overline{x}_{ii} = x_{ii}$ for $i \notin W$;

c) $\overline{x}_{ii} > x_{kd}$ for all $(k, d) \neq (j, j)$, $j \in W$;

d) $\overline{x}_{ii} = \overline{x}_{jj}$ for $i, j \in W$ if and only if $\lambda_i(X)$ and $\lambda_j(X)$ have equal composition;

e) $\overline{x}_{ii} > \overline{x}_{jj}$ for $i, j \in W$ if and only if $\lambda_i(X) > \lambda_j(X)$ (comparison in the sense of composition of matrices);

f) Cf. M 2.1f.

Set

$$\sigma_{1,W}(X) = \text{Stab } \overline{X}$$

Remark. Even if $\dim \sigma_{1,W}(X) = \dim X$, it is possible that $\sigma_{1,W}(X) \neq X$. However, $\sigma_{1,W}(X) \sim X$ in this case.

4.2. Lemma. a) The algorithm $X \longrightarrow \sigma_{1,W}(X)$ is invariant, i.e., $\sigma_{1,W}(\tau X \tau^{-1}) = \tau(\sigma_{1,W}(X))\tau^{-1}$ for $\tau \in \text{Sym } V(X)$;

b) $\text{Aut } X = \text{Aut } \sigma_{1,W}(X)$.

Proof. (a) is evident; (b) follows from (a).

4.3. Lemma. If $X \sim \sigma_{1,W}(X)$ and $\lambda_i(X) = \sum_{j\in I_i} Y_j f_j^{(i)}$, then $I_i = I_j$ for all $i, j \in W$ and the identity map $I_i \longrightarrow I_j$ is a natural weak equivalency.

Proof. The first assertion follows directly from 4.1d,e. The second one follows from the first and from the fact that the $\lambda_i(X)$ are obtained by simultaneous stabilization (cf. M 4.4).

4.4. Suppose that the entries of the graph X belong to a linearly ordered set of variables. Assume $\sigma_{1,W}(X) \sim X$. Let $\lambda_k(X) = \sum_{i=1}^{n} y_i f_i^{(k)}$. Put

$$\tilde{f}_i = \sum_{k \in W} f_i^{(k)}$$

Let $\tilde{f}_i = \sum_{j=1}^{m_i} a_{ij} g_{ij}$, where $a_{ij} \in \mathbf{Z}$, $a_{ij} > 0$, $a_{ij} > a_{ij+1}$ and where g_{ij} are disjoint (0, 1)-matrices. (That is, g_{ij} has ones at those positions where $\tilde{f}_i$ has a_{ij} and g_{ij} has zeroes otherwise.) Set (cf. M 2.2)

$$\sigma_{2,W}(X) = \text{Stab } X(g_{11} \cap \ldots \cap g_{1m_1} \cap \ldots \cap g_{n1} \cap \ldots \cap g_{nm_n})$$

4.5. <u>Lemma</u>. Suppose that the entries of X are linearly ordered and $X \sim \sigma_{1,W}(X)$.

a) The algorithm $X \longrightarrow \sigma_{2,W}(X)$ is invariant;

b) $\text{Aut } X = \text{Aut } \sigma_{2,W}(X)$.

<u>Proof</u>. See 4.2.

4.6. <u>Remark</u>. The stability with respect to $\sigma_{1,W}$, $\sigma_{2,W}$ is usually sufficient to prove theorems. Actually we use only Theorem 4.7 below. We give, however, in 4.9, 4.10, some additional operations. Geometrically, all these operations of stabilization can be described as follows. Each set $V(X_{ii})$ and each graph e_j fall into pieces in each $\lambda_m(X)$. If there is a difference in the coloration of these pieces for different m, then it gives rise to a difference of the corresponding vertices. They should, therefore, be repainted in different colors (this is $\sigma_{1,W}$). If different pieces of e_j behave differently with respect to the family $\{\lambda_m(X)\}$, we can repaint edges of e_j (this is $\sigma_{2,W}$). And so forth.

4.7. <u>Theorem</u>. Let $X = (X_{ij})$ be a stationary graph with linearly ordered entries,

$$W = V(X_{tt}),\ X \sim \sigma_{1,W}(X) \sim \sigma_{2,W}(X)$$

Let $X = \sum_{i=1}^{n} x_i e_i$, $\lambda_m(X) = Y^m = \sum_{i=1}^{N} y_i f_i^{(m)} = (Y^m_{ij})$. Then

a) If $f_i^{(m)} \cap e_r \neq 0$, then $\sum_{m\in W} f_i^{(m)} = a_i e_r$ where a_i is the product of $|W|$ by the number of ones in $f_i^{(m)}$ (for any m) divided by the number of ones in e_r.

b) Let $e_i \subset X_{pt}$, $d = d(e_i)$, and let $r_{1,m}, \ldots, r_{d,m}$ be the numbers of those positions of the m-th row of e_i where ones stand. Then there exists j such that for all $m \in W$ one has $V(Y^m_{jj}) = \{r_{1,m}, \ldots, r_{d,m}\}$. In particular, $|Y^m_{jj}| = d$.

c) $\operatorname{rg} Y^m = \sum_i \dim X_{it}$.

Proof. By the definition of the simultaneous stabilization, $f_i^{(m)} \cap e_r \neq 0$ implies $f_i^{(q)} \subset e_r$ for all $q \in W$. By stability with respect to $\sigma_{2,W}$, $\tilde{f}_i = a_i e_r$.

The equality between a_i and the number asserted in Part (a) of the theorem is easily verified.

Let us prove (b). Define s by the condition $Y^m_{ss} = (y_{mm})$. Take $f_q^{(m)} \subset Y^m_{sj}$ and assume $f_q^{(m)} \cap e_i \neq 0$. Then $\sum_{m\in W} f_q^{(m)} = a_q e_i$.

However, every non-zero entry of every matrix $f_q^{(m)}$ is contained in the m-th row. By the condition $f_q^{(m)} \subseteq e_i$ (which follows from $f_q^{(m)} \cap e_i \neq 0$) one has $d(f_q^{(m)}) \leq d(e_i) = d$. Now (a) implies that $a_q \leq 1$, hence $a_q = 1$. But then $d(f_q^{(m)}) = d(e_i)$, that is $|Y^m_{jj}| = d(f_q^{(m)}) = d(e_i)$.

Let us now deduce (c) from (b). By (b) for any $e_i \subset \bigcup_k X_{kt}$, there exists $j = j(i)$ satisfying the conditions of (b). In particular, the equality $j(i) = j(r)$ implies $i = r$. Since $\Sigma|Y^m_{jj}| = \sum_{e_i \subset \bigcup_k X_{kt}} d(e_r) = |X|$, (c) follows.

4.8. Corollary. Under the conditions of 4.7, the blocks Y^m_{ii} of the graph $\lambda_m(X)$ can be numbered by the numbers of those e_i which lie in $\bigcup_k X_{kt}$.

Proof follows directly from (b) and (c), cf. also the end of the proof of Theorem 4.7.

4.9. Suppose that the variables of a stationary graph X are linearly ordered and

$X \sim \sigma_{1,W}(X) \sim \sigma_{2,W}(X)$. Let

$$\{\lambda_{ij}(X)\} = \mathrm{Stab}\{\lambda_i(X) \cap \lambda_j X)\}$$

(simultaneous stabilization). Define the graph $\overline{X} = (\overline{x}_{ij})$ in the following manner.

a) $\overline{x}_{ij} = x_{ij}$ for $(i, j) \notin W \times W$;

b) $\overline{x}_{ii} = x_{ii}$ for all i;

c) $\overline{x}_{ij} > \overline{x}_{kt}$ if $(i, j) \in W \times W$, $(k, t) \notin W \times W$;

d) $\overline{x}_{ij} = \overline{x}_{kt}$ for $i, j, k, t \in W$, $i \neq j$, $k \neq t$, if and only if the pair (x_{ij}, composition of $\lambda_{ij}(X)$) coincides with the pair (x_{kt}, composition of $\lambda_{kt}(X)$).

e) $\overline{x}_{ij} > \overline{x}_{kt}$, for $i, j, k, t \in W$, $i \neq j$, $k \neq t$, if and only if the pair (x_{ij}, composition of $\lambda_{ij}(X)$) > the pair (x_{kt}, composition of $\lambda_{kt}(X)$).

Set $\sigma_{3,W}(X) = \mathrm{Stab}\ \overline{X}$.

4.10. Lemma. Suppose that $X \sim \sigma_{1,W}(X) \sim \sigma_{2,W}(X)$. Then

a) The algorithm $X \longrightarrow \sigma_{3,W}(X)$ is invariant.

b) $\mathrm{Aut}\ X = \mathrm{Aut}\ \sigma_{3,W}(X)$.

Proof. Evident.

4.11. Suppose that the variables of a stationary graph X are linearly ordered and that $X \sim \sigma_{1,W}(X) \sim \sigma_{2,W}(X) \sim \sigma_{3,W}(X)$. Let $\lambda_{pq}(X) = \Sigma z_i f_i^{(p,q)}$. Take $e_r \subset X_{tt}$ (recall $W = V(X_{tt})$). Set

$$\mathcal{F}_{i,r} = \sum_{(p,q) \text{ an edge of } e_r} f_i^{(p,q)}$$

Let

$$\widetilde{f}_{i,r} = \sum_{j=1}^{m(i,r)} a_{irj}\, g_{irj}$$

where $a_{irj} > a_{irj+1}$, $a_{irj} > 0$, g_{irj} - (0, 1)-matrix. Set

$$\sigma_{4,W}(X) = \mathrm{Stab}\ X(g_{111} \cap \dots \cap g_{11m(1,1)} \cap \dots \quad \dots \cap g_{n,d,m(n,d)})$$

where $d = \dim X_{tt}$.

4.12. Lemma. Let $X \sim \sigma_{1,W}(X) \sim \sigma_{2,W}(X) \sim \sigma_{3,W}(X)$.

a) The algorithm $X \longrightarrow \sigma_{4,W}(X)$ is defined invariantly.

b) $\mathrm{Aut}\ X = \mathrm{Aut}\ \sigma_{4,W}(X)$.

Proof. Evident.

4.13. Definitions. Let X be a stationary graph whose variables are linearly ordered. We say that X is stable of depth 1 (or simply, X has depth 1) with respect to $W = V(X_{tt})$, if

$$X \sim \sigma_{1,W}(X) \sim \sigma_{2,W}(X) \sim \sigma_{3,W}(X) \sim \sigma_{4,W}(X)$$

We say that X has depth 1 with respect to $W = \bigcup_{t \in J} V(X_{tt})$ if X has depth 1 with respect to every $V(X_{tt})$, $t \in J$. We say that X has depth 1 if it has depth 1 with respect to $V(X)$.

5. Comments on the definition of stabilization

5.1. For an arbitrary daughter system $D_W(X) = \{Y^1, \dots, Y^m\}$, let us set $\widetilde{D}_W = \mathrm{Stab}\{Y^i\} = \{\widetilde{Y}_i\}$. Then the operations $\sigma_{i,W}$ can be easily defined (one should substitute $\lambda_i(X)$ by $\widetilde{Y}^i$ in the corresponding definitions). The Lemmas 4.2, 4.5, 4.10, 4.12 hold if the system $D_W(X)$ is defined invariantly.

5.2. The operations $\sigma_{i,W}$ can be complemented by an operation which is a hybrid of stabilization with respect to the kernel, and of the operation $\sigma_{I,W}$. Namely, for every $i \in W$, one can consider the numerical invariants $\mu_{3,\Pi}(\varkappa_{W_i}(\lambda_i(X)))$ (cf. E 6.2). Analogously one can consider

$$\mu_{3,\Pi}(\varkappa_{W_i} \cap W_j(\lambda_{ij}(X)))$$

etc. Here W_i are invariantly defined subsets of $V(\lambda_i(X))$.

6. Depth > 1 and variants of definition of depth

In this subsection we give only definitions or sketches of definitions.

6.1. Let us say that X has depth $(m + 1)$ if $\lambda_i(X)$ has depth m for all $i \in V(X)$.

6.2. The depth m can be defined via the consideration of the system of graphs $\lambda_i(\lambda_j(\dots \lambda_t(X) \dots))$, where $|\{i, j, \dots, t\}| = m$, $i, j, \dots, t$ lie in invariantly defined subsets and where stabilization is simultaneous for all sets of those graphs.

6.3. One can say that X has depth m if the number of graphs belonging to any given isomorphism class of graphs of degree $\leq m$ and containing a given edge x_{ij}, depends only on the isomorphism class and the "color" of x_{ij}.

In this sense a stationary graph has depth 3. A variant of this definition and arising properties are discussed in Section AD.

6.4. Let $X = \sum_{i \in I} x_i e_i$ be a stationary graph. Define the "dual" graph $\hat{X}$ in the following manner.

a) $\hat{X} = (\hat{X}_{ij})_{i,j \in I}$;

b) $V(\hat{X}_{ii})$ is the set of the edges of the graph e_i;

c) If a is an edge of e_i and b is an edge of e_j, then the "color" of the edge (a,b) of $\hat{X}$ is the "color" of the triangle constructed on the vertices of a and b if a and b have a common vertex and the set of the "colors" of the quad-rangle constructed on the vertices of a and b if they have no common vertex.

These conditions permit one to fill the entire matrix $\hat{X}$ by variables depending on the type of relation of the edges.

Now consider Stab $\hat{X}$ and say that X has <u>depth</u> (or height?) 1 if blocks $\hat{X}_{ii}$ do not decompose in the graph Stab $\hat{X}$. If, however, they decompose, the differences among the edges of graphs e_i are revealed, and the dimension of X can be invariantly increased (as in 4.1).

The graph $\hat{X}$ is perhaps an interesting object. However, no results are known to us about this graph, and we, therefore, proceed without stopping.

7. <u>Examples</u>. It is difficult to give detailed examples, where the procedures described above really work. Indeed, such examples would be first encountered among graphs with 25 vertices and, therefore, would be very complex.

We shall give partial examples using the graph from 26-family. Partial means that we shall not compute $\sigma_{i,W}$, but we shall only show that the result of $\sigma_{1,W}$ can be different from the result of Stab and that it can give a partition into orbits of the automorphism group even when X is a cell.

7.1. A common assumption in the examples given below is that for the neighbor graphs Γ_i of the 26-family (given by the pictures in Section U), the stationary graphs Stab $X(\Gamma_i)$ are different (have different structure constants a_{ij}^k). We shall not check this here.

7.2. Our approach is as follows. Let e be the matrix #i from the 26-family (cf. Section U). Consider the graph

$$X = x\, E_{26} + y\, e + z(\widetilde{I}_{26} - e)$$

Instead of the graphs $\lambda_j(X)$, $j = 1, \ldots, 26$, we shall consider only the neighbor graph in e of the vertex j. The isomorphism class of this neighbor graph is given in the column "TYPE" on the same table as e itself. By the assumption stated in 7.1, this information is sufficient to give (with the help of $\sigma_{1,W}$) the partition of

V(X). We shall refine this partition using some special properties of neighbor graphs. Our aim is to achieve partition into orbits of Aut e (this partition is given on the same page as e itself).

7.3 Let us take $i = 3$ (the 26-family, #3). the vertices according to the type of their neighbor graphs fall into the following groups:

(1,2,3), (4,7,13,18), (5,6,12), (8,9,14,15,21,22,23,24,25), (10,11,16,17,19,20), (26).

In particular, we infer that in $\sigma_{1,W}(X) = (Y_{ij})$ we have $\{26\} = V(Y_{ii})$ for some i. Now consider the neighbor graph of the 26-th vertex. It is of type 2. In this graph only vertex 5 (in canonical numeration) is not contained in any triangle. Therefore, the fifth vertex of the neighbor graph of the 26-th vertex is separated. So we have separated 4-th vertex (that is, $V(Y_{jj}) = \{4\}$ for some j).

In the neighbor graph of the 4-th vertex (which is of type 3) only the vertices (1,2,3,23,24,25) are contained in triangles. This partition together with the partition given above (corresponding to the types of the neighbor graphs) gives us (as the intersection) the partition into the orbits of Aut e.

7.4. Let us now take $i = 9$ (the 26-family, #9). This case is somewhat more difficult because there are only 4 types of neighbors, and because the result we want to achieve is the split graph. Only vertex 1 has the neighbor graph of type 7. Hence in $\sigma_{1,W}(X) = (Y_{ij})$ we have $\{7\} = V(Y_{jj})$ for some j. Therefore, in addition to the partition of the vertices according to the neighbor types, we get the partition

$$\{1\},\ [2,11],\ [12,26]$$

The intersection of this partition with the partition according to neighbor types gives us the partition

(1), (2,6,7,11), (3,4,5,8,9,10), (12,18,25), (13,15,17,20,22,24,26)
(14,16,19,21,23) (*)

Consider the vertices 12, 18, 25. Their neighbor graphs contain the vertices 2÷11 with the following multiplicity

(7,10) -- with multiplicity 0

(2,4,8,11) -- with multiplicity 1 (**)

(3,5,6,9) -- with multiplicity 2

Since $\{1\} = V(Y_{jj})$ for some j, it follows that the three sets above are unions of some $V(Y_{ss})$.

Intersecting the sets of (*) and (**), we get the following partition

(1), (2,11), (3,5,9), (4,8), (6), (7), (10), (12,18,25)
(13,15,17,20,22,24,26), (14,16,19,21,23) (***)

This shows that the intersection of this partition with the neighbor sets of vertices 6, 7, 10 is invariantly defined. These sets are respectively

(1,3,4,10,16,18,19,24,25,26)

(1,3,4,11,17,20,21,22,23,26)

(1,6,8, 9,13,14,17,21,24,26)

The intersection of these sets with (***) gives us the partition

(1), (2), (3), (4), (5), (6), (7), (8), (9), (10), (11), (12), (13), (14), (15), (16,19), (17,20), (18,25), (21), (22), (23), (24), (26)

To split the remaining three pairs

(16,19), (17,20), (18,25)

note that the neighbor graph of the vertex 2 contains vertices 16 and 17 but does not contain vertices 19, 20. It splits the first two pairs. The neighbor graph of the vertex 4 contains vertex 25 but does not contain vertex 18. This concludes the splitting.

P. EXAMPLES OF RESULTS USING THE STABILITY OF DEPTH 1.

1. Statements and proofs of theorems given below make use of the notions introduced in the preceding Sections. The theorems themselves are analogues of some simple results of permutation group theory. This implies that possibly deeper results of that theory can also be restated and reproved in the setting of cellular algebras.

2. Theorem 2.1, below, is used in the algorithm of Section R. In fact, this theorem is a justification of the approach taken in that algorithm.

In this Section, X stands for a stationary graph with linearly ordered variables.

2.1. <u>Theorem</u>. Let $X = (X_{ij})$ be a stationary graph of depth 1 with respect to $W = V(X_{11})$, $|W| = n$. If $\lambda_i(X) = R$ for $i \in W$, then

a) $\mathfrak{A}(X_{11}) \simeq \mathbb{Z}[G]$, where G is a group of order n;

b) $\mathrm{Aut}\, X \simeq G$;

c) the orbits of G are the sets $V(X_{ii})$.

<u>Proof</u>. Since $\lambda_i(X) = R$, one has by Theorem O4.7b, $d(e_m) = 1$ for any $e_m \subset X_{11}$. Hence (a) follows from G 1.

Let us now use Theorem O 3.5. Note that one has (in the notations of that theorem) $W_1 = W$, since X has depth 1 with respect to W. Hence Theorem O 3.5 and (a) above yield (b).

Let us prove (c). Again, by condition $\lambda_i(X) = R$, and by Theorem O 4.7b, one has $d(e_m) = 1$ for $e_m \subset X_{1t}$. Fix m so that $e_m \subset X_{1t}$. Then e_m defines (cf. I 3) X_{tt} as the factorgraph of X_{11}. Let $V_1, \ldots, V_r$ be the imprimitivity system for X_{11} defined by e_m. The action of G on $V(X_{11})$ induces the transitive action of G on the sets V_i and, consequently, on $V(X_{tt})$. By (b), G acts as an automorphism group, hence (c) is proved.

<u>Remark</u>. Actually, the fact that X_{tt} is a factor of X_{11} permits one to identify

$V(X_{tt})$ with G/H for some subgroup H of G, and the action of G on both sets coincides.

2.2. Proposition. Let X be a primitive cell of depth 1, $Y^q = \lambda_q(X) = (Y^q_{ij})$. Then either $|Y^q_{ii}| > 1$ if $V(Y^q_{ii}) \neq \{q\}$ or $\mathfrak{a}(X) = \mathbb{Z}[Z_p]$, p a prime number.

Proof. If $|Y^q_{ii}| = 1$, let e_i be the corresponding basic graph, according to O 4.7, 4.8. Then $d(e_i) = 1$. Our assertion now follows from K 1.

3. Theorem. Let $X = (X_{ij})$ be a stationary graph of depth 1 with respect to $W = V(X_{11})$. Assume that there exists a basic graph $e_m \subset X_{12}$ such that $d(e_m) = 2$. Then there exists a non-oriented (i.e., simple) graph $e_t \subset X_{22}$ and a normal subcell $\mathcal{L}$ in X_{11}, such that $X_{11}/\mathcal{L}$ contains the basic graphs e_j, $j \in J$, whose sum $\sum_{j \in J} e_j$ is isomorphic to the edge graph of graph e_t. In particular,
$$|X_{11}/\mathcal{L}| = \frac{|X_{22}| \cdot d(e_t)}{2}.$$

Proof. Put $Y^i = \lambda_i(X) = (Y^i_{kl})$, $i \in W$. Since $e_m \subset X_{12}$, X has depth 1 with respect to W, and $d(e_m) = 2$, it can be assumed (cf. O 4.7, 4.8) that $|Y^i_{mm}| = 2$, $Y^i_{mm} \subset X_{22}$ and $V(Y^i_{mm})$ is defined by e_i in the manner described in O 4.7, 4.8. We have $Y^i_{mm} = x\, E_2 + y\, \tilde{I}_2$. Then there evidently exists $e_t \subset X_{22}$ such that $e_t \cap Y^i_{mm} = \tilde{I}_2$. It is also clear that e_t is unique and that $e_t' = e_t$. Since $e_t \cap Y^i_{mm} = \tilde{I}_2$, i defines an edge of the graph e_t. Let D be the set of the edges of the graph e_t. We defined the map $\psi : V(X_{11}) \longrightarrow D$. Let us show that it is surjective. Let $Y^i_{mm} = x\, f_1^{(i)} + y\, f_2^{(i)}$, where $f_2^{(i)} = \tilde{I}_2$. Since X has depth 1, we have $\sum_i f_2^{(i)} = a\, e_t$ which is equivalent to surjectivity.

Now let $\mathcal{L}$ be the normal subcell of X_{11} defined by equality of the rows of the matrix e_m (cf. J 2). Note that the corresponding imprimitivity system coincides with $\{\psi^{-1}(d)\}_{d \in D}$. Thus we have the equality $|X_{11}/\mathcal{L}| = |D| = |X_{22}| \cdot d(e_t) \cdot 2^{-1}$. Consider the factor-graph $\bar{X} = (\bar{X}_{ij})$ of X by the system of normal subcells $\{\mathcal{L} \triangleleft X_{11},\ 1 \subset X_{ii},\ i > 1\}$. If $\bar{e}_m$ is the image of e_m in X_{12}, then $d(\bar{e}_m) = 2$. Hence we can assume that $\bar{X} = X$, $\mathcal{L} = 1$, and $\bar{e}_m = e_m$. We have $e_m \cdot e_m' \subset X_{11}$, $e_m \cdot e_m' = 2e_1 + \sum_{k \neq 1} a^k_{mm'}\, e_k$, where e_1 is the unity of the cell $\mathfrak{O}(X_{11})$. Since $\mathcal{L} = 1$, no pair of the rows of e_m coincides. Hence $a^k_{mm'} = 0$ or

1 for all $k \neq 1$. Put $J = \{k : a^k_{mm'} = 1\}$. Let us show that $\tilde{e} = \sum_{j \in J} e_j$ is the edge graph of e_t. Take $p, q \in W$. If $\psi(p)$ and $\psi(q)$ have a common vertex, then the rows of matrix e_m with numbers p and q have ones in the same column. This means that the edge (p, q) is an edge of the graph $\tilde{e}$. The converse assertion is proved identically.

4. (Compare [Wi 1, 17.7], [Qu 1]), [Ca 1])

4.1. Let X be a primitive cell of depth 1, $X = \Sigma x_i e_i$, $n_i = d(e_i)$, $e_i e_j = \Sigma a^k_{ij} e_k$, $Y^j = \lambda_j(X) = (Y^j_{st}) = \Sigma y_i f^j_i$. Suppose that the diagonal blocks of Y^j are numbered according to 0 4.7, 4.8 by numbers of those e_i which split them off. Let $Y^j_{11} = S_m$, $m > 2$.

4.2. Let $Y^j_{11} = x E_m + y \tilde{I}_m$ and take $q(j)$ so that $e_{q(j)} \cap Y^j_{11} = \tilde{I}_m$. Since X has depth 1 (and is, in particular, stable under $\sigma_{1,W}$), one has $q(j) = q(k)$ for all $j, k \in V(X)$. Hence there exists a unique q such that for all $j \in W$ one has $e_q \cap Y^j_{11} = \tilde{I}_m$

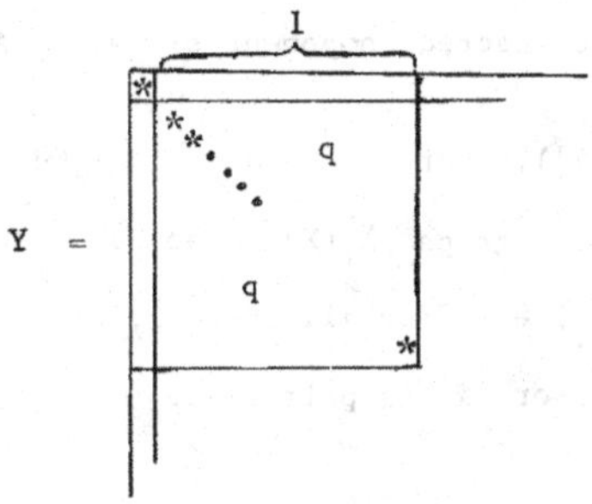

4.3. Theorem.

$$q = q'$$

$$a^1_{1q} = m - 1, \quad a^q_{qq} \geq m - 2$$

$$n_q \text{ divides } m(m - 1)$$

$$n_q > m$$

Proof. Since $\tilde{I}_m = \tilde{I}'_m$, we have $e_q = e'_q$, i.e., $q = q'$. From the Figure in 4.2, one has $a^{1}_{1q} = m - 1$, $a^{q}_{qq} \geq m - 2$. Since (cf. D 4 c 8)

$$m(m-1) = n_1 \, a^{1}_{1q} = n_q \, a^{q}_{1'1}$$

we see that n_q divides $m(m-1)$. If $n_q \leq m - 1$, then $a^{q}_{1'1} \geq m$, that is $a^{q}_{1'1} = m$. By K3 this implies imprimitivity of X, which contradicts our assumptions. Hence $n_q \geq m$. Suppose $n_q = m$.

Let Γ_i be a complete subgraph of e_q, defined by the condition $\Gamma_i \cap Y^{i}_{11} = \tilde{I}_m$. Then every edge of e_q is contained in $a = m(m-1)n_q^{-1}$ graphs Γ_i, and every vertex is contained in the m graphs Γ_i. Let $V_i = V(\Gamma_i)$. Let $V_1, \ldots, V_m$ contain $t \in V(X)$. Since $a \geq 1$, one has $|\bigcup_{i=1}^{m} V_i| = n_q + 1$ (= the neighbors of t and t itself). If $n_q = m$, then $|\bigcup V_i| = m + 1$. Since $V_i \neq V_j$ for $i \neq j$ (primitivity), it follows that $e_q|_{\bigcup_{i=1}^{m} V_i}$ is a complete graph with $(m+1)$-vertices, that is, it is a connected component of e_q. A contradiction with H 12.

5. Theorem (cf. [Wi 1, 10.4]). Let X be a cell of depth 1 and suppose that for all $i \in V(X)$, the stationary graph $\lambda_i(X)$ also has depth 1. Let $X = \sum_{i=0}^{d} x_i e_i$, $e_0 = E_n$, $d(e_i) = m$ for all $i \in [1, d]$. Then either $\lambda_i(\lambda_j(X))$ is split for all $i, j \in V(X)$, or X is primitive.

Proof is given in a series of steps. Suppose that X is imprimitive. Let $\mathcal{L}$ be a normal subcell in X, and $V_1, \ldots, V_b$ be the corresponding imprimitivity system. Set $|V_i| = r$, $X = (X_{ij})$, $V(X_{ii}) = V_i$. Let s_t be the t-th row of X. Let e_i, $i \in J$, define our normal subcell, that is, $e_i \in \mathcal{L}$ if and only if $i \in J$. Put $|J| = a + 1$. Evidently, $0 \in J$.

We shall show that our assumption leads us into the first case of the alternative.

5.1. $(m, r) = 1$.

Proof. $r = |V_1| = \sum_{i \in J} d(e_i) = 1 + am$.

5.2. $|e_j \cap s_q \cap X_{kt}| \leq 1$ if $k \neq t$.

Proof. If $q \notin V_k$, then the intersection is empty. Hence, it may be assumed that $q \in V_k$. Set $s_{q,t} = s_q \cap X_{kt}$. Let

$$I_{q,t} = \{i : e_i \cap s_{q,t} \neq 0\}, \ \Gamma = \sum_{i \in I_{q,t}} e_i$$

Let T be the set of neighbors of the vertex q in the graph Γ. We have $T = \bigcup V(X_{ii})$, where the sum is taken over those j for which $I_{q,j} = I_{q,t}$ (by similarity of rows, cf. I 1.1). Hence, r divides T. On the other hand, $|T| = |I_{q,t}|$. By 5.1 it follows that r divides $|I_{q,t}|$. Since $|I_{q,t}| \leq r$, we have $|I_{q,t}| = r$, whence our assertion.

5.3. If $p \in V_i$, $q \in V_j$, $i \neq j$, then

$$\lambda_p(\lambda_q(X)) \sim \varkappa_{V_i \cup V_j}(\lambda_p(\lambda_q(X))$$

Proof. Set $Y = \lambda_p(\lambda_q(X))$, $Z = \lambda_q(\lambda_p(X))$. Since X is imprimitive, it follows that the blocks with the numbers from 1 to a of the central decomposition of the stationary graph $\lambda_q(X)$ all lie in X_{jj} (we suppose here that they are numbered according to 0 4.7, 4.8). The above remark and 5.2 imply that deletion of $p \in V_i$, $i \neq j$, splits $Y|_{V_j}$, that is, $Y \sim \varkappa_{V_j}(Y)$. Analogously, $Z \sim \varkappa_{V_i}(Z)$. Since evidently $Y \sim Z$, our assertion follows.

5.4. $\lambda_p(\lambda_q(X)) = R$ for all $p, q \in V(X)$.

Proof. Let $q \in V_i$. Set $Y^q = \lambda_q(X) = (Y^q_{ij}) = \Sigma\, y_i\, f^{(q)}_i$. We have $V_i = q \cup \bigcup_{s=1}^{a} V(Y^q_{ss})$. By 0 4.7, we have $|Y^q_{ss}| = m$ for all $s \neq 0$. Let $p \in V_j$, $j \neq i$. Then $\lambda_p \lambda_q$ splits V_i.

Let $p \in V(Y^q_{tt})$. Since Y^q has depth 1, then by 0 4.7, $d(f_k) = 1$ for all $f_k \subset Y^q_{ts}$, $1 \leq s \leq a$. Since $|Y^q_{tt}| = |Y^q_{ss}| = m$, it follows that $d(f'_k) = 1$. Consider now $Y^q_{ts} \cdot Y^q_{st} \subset Y^q_{tt}$. It follows from above that $d(f_k) = 1$ for all $f_k \subset Y^q_{tt}$, i.e., $Y^q_{tt} = \mathbf{Z}[G]$. Now 5.2 and the condition $d(f_k) = d(f'_k) = 1$ for all $f_k \subset Y^q_{ts}$, $s \leq a$, imply that $Y^q_{ss} = \mathbf{Z}[G]$ for all $s \leq a$. Since t was taken arbitrary from numbers greater than a (to satisfy $V(Y^q_{tt}) \cap V_i = \emptyset$), we have $Y^q_{tt} = \mathbf{Z}[G]$ for all $t \neq 0$ and $Y^q_{tt} \cong Y^q_{11}$ for all $t \neq 0$. Therefore, $d(f_k) = d(f'_k) = 1$ for all k.

Thus our assertion, and the theorem, are proved.

Q. SOME DEFINITIONS AND EXPLANATIONS ABOUT EXHAUSTIVE SEARCH.

1. Below we give some definitions related to exhaustive search. We do this in order to construct a frame of reference for subsequent Sections.

Descriptions of algorithms are usually omitted; if these algorithms are sufficiently complicated, use very ambiguous or, on the contrary, very formal (e.g., ALGOL) language. We tried to take the middle road. So we stopped at some distance from complete strictness (and senselessness).

It seems that the formalism proposed below is suitable for the description of some exhaustive methods. It was used, in particular, by G. M. Adelson-Velsky, V. L. Arlazarov and M. V. Donskoy to prove optimality of the branch-and-bound method and to describe in a more exact language new developments in their chess program (which, it should be reminded, won world chess programs' competition in 1975).

2. Let us first give a very approximate and down-to-earth description of the notions involved.

First of all exhaustive search is a method used to solve problems of the following kind. We are given a finite set V and we are required to find one or several elements of V satisfying certain conditions.

2.1. If elements of V are given explicitly, then one checks every one of them in turn for the required property.

2.2. But usually the situation is more complicated. Namely, usually we are given rules for the construction of some subsets, say $V_1, \ldots, V_m$ of V, and for the subset V_i we are given rules for the construction of its subsets $V_{1i}, \ldots, V_{d(i)i}$, and so forth. The elements of V will appear as one-point subsets somewhere far down the line.

The rules which are used to construct subsets may depend and, as a matter of fact, sometimes do depend on the set to which they are applied.

2.3. It is customary to associate with the above sequence of subsets an oriented graph. The set of vertices of this graph is the set of all subsets of V, which were constructed by the application of the rules. So the sets V, V_i, V_{ij}, etc., are vertices. Vertex a is joined to vertex b, if the subset corresponding to b is obtained from the subset corresponding to a by the application of the given rules. V represents the root of this graph.

If every subset is constructed at most once, the resulting graph is a tree. This happens for example in the case when the application of the rules so any subset generates disjoint subsets.

2.4. An exhaustive search is described by the order in which we consider the vertices of the graph described above. If our search brings us to some vertex of this graph, the length of the path from the root to the vertex under consideration is called the depth (or the level) of our search at this moment. This definition depends on the path which leads from the root to the given vertex. If there is only one such path (that is, if the graph is a tree), the depth of the search depends only on the vertex.

2.5. The usual order of the search is called "depth-first search". In this search one goes down to the end point, say a_1, of some path say $a_n, \ldots, a_1$. If this end point is a solution of our problem, the search is finished. If it is not, the search takes in turn all successors of a_2, then a new successor of a_3, say a_2', and considers in turn its successors. Et cetera until a solution is met or its absence is established.

In this way the required storage space is of the order of the maximal length of a path in our graph. (We have to remember the whole sequence of subsets as well as the information about the next successor for every one of these subsets.)

An antipode to the depth-first search is "breadth-first search". In this search one first constructs all vertices of the first level, then all vertices of the second level and so forth. In this case, generally speaking, we have to use storage space of the order of the number of vertices of the given level. However it is possible that, having that much information, one would be able to establish that some of the subsets of the given level do not contain the searched-for points of V and therefore can be rejected (cf. 2.7, 3.5). If this does not happen, the breadth-first search would fail, owing to the lack of storage (which is even more scarce than time).

The algorithm of R8.1 is a breadth-first search (it is not meant to be programmed) and the algorithms of Sections S, T are depth-first searches.

2.6. A rough estimate of the "time" to be consumed by an exhaustive search can be obtained in the assumption that the application of the rules to every subset uses the same amount of time. Then the general amount of time is a multiple of the number of vertices we searched through.

2.7. In many cases there exist (and sometimes they indeed are known) means to establish the absence of elements with required properties. The application of the corresponding criteria is called variant rejection or cut off.

2.8. Frequently (cf., 6.2) there are several ways to associate an exhaustive search to a given problem. In this case the choice of an exhaustive search affects the possibilities for variant rejections. It is natural to organize an exhaustive search in such a way, that the number of vertices searched through would be as small as possible.

In other problems rules are given explicitly (cf. 6.1).

2.9. The time required for a search can also be saved by a clever choice of the order in which vertices are searched.

2.10. The sequence of vertices of a search graph is called a forced variant, if the application of the rules to the corresponding sets gives rise to subsets of which at most one is not subject to variant rejection. This means that during the search we have to move only in one direction from the vertices of the forced variant. Examples of forced variants are in T2.4, T3.4.

2.11. Speaking about variant rejections and the choice of an order of a search, we have to keep in mind that it costs (computer) time and (storage) space to implement sophisticated procedures. The price of each verification for a possibility of a variant rejection can be high (for example, it can involve some exhaustive search in itself), but the number of rejected sets can be small. In this case the use of such a variant rejection would be wasteful. Similarly a complicated choice of an order of a search ("What will be my next step?") can be improper.

Often we do not know the price of an application of the corresponding decision procedures. In this case the success of their use depends on the ability to do a rough experiment, and on good luck. Such procedures, which will hopefully lead to a speedup (but one does not know for certain whether they will have this or the opposite effect) we call <u>heuristics</u> (examples are T2.5, S3.4).

Sometimes considerations leading to powerful variant rejections, but possibly to a wrong result (if a child was thrown away along with a bath), are also called <u>heuristics</u>. We do not use this word in this latter sense in this volume.

3. <u>A formalization.</u>

3.1. A description of a problem.

For a finite set V let $P(V)$ denote the set of all subsets of V (in particular, $|P(V)| = 2^{|V|}$). We identify V with the subset of $P(V)$ consisting of all one-element subsets of V.

Suppose we are given a computable function $F : P(V) \to \mathbb{N}$. It will be called an <u>estimate function</u>.

Problem: Find a subset $U \subseteq V$ such that F is defined on U and attains its maximum value on U.

3.2. Exhaustion.

Define a function of an exhaustion as a computable map

$$f : P(V) \to P(P(V))$$

satisfying the following conditions

a) If $f(U)$ is defined for $U \in P(V)$, then $F(U)$ is defined;

b) If $f(U)$ is defined for $U \in P(V)$, then $f(U) \in P(P(U))$;

c) At least one subset U, which is a solution of our problem, belongs to the image of $f^i(V)$ for an appropriate i.

3.3. The graph of an exhaustion.

The graph T of the mapping f is called the graph of exhaustion f.

More explicitly

a) the vertices of T are elements of $P(V)$ where f is defined and which belong to $\bigcup_{i \geq 1} f^i(V)$.

b) There is an edge from $a \in V(T)$ to $b \in V(T)$ if and only if $a \in f(b)$.

(Here and below we identify vertices of T with elements of $P(V)$).

Then $\{V\} \in P(V)$ is the root of T. Let $T(m)$ be the set of the vertices of T which are at a distance m from the root of T. If $U \in V(T)$ let

$$Q_U = f(U)$$

be the set of all successors of U, and

$$P_U = \{M \in P(V) \mid U \in f(M)\}$$

the set of all predecessors of U. Let $T(U)$ be the graph of exhaustion $f|P(U)$ (it is the subgraph hanging at $U \in V(T)$).

Finally, let End T = End f be the set of all terminal points of T, that is, the set of all $U \in P(V)$ such that $f(U)$ is not defined, and Solv T = Solv f the set of all $U \in P(V)$ which are solutions of our problem.

3.4. Search on the graph of an exhaustion. A computable function

$$\varphi : \mathbb{N} \to V(T)$$

is called a search if

a) $\bigcup_{n \in \mathbb{N}} \varphi(n)$ contains a point from Solv T;

b) $\varphi(n+1) \in \bigcup_{i \leq n} \varphi(n)$

An exhaustive search is therefore a triple (T, φ, F), where T is the tree of a search, φ is a search over T, and F is an estimate function.

3.5. Variant rejections.

Let (T, φ, F) be an exhaustive search. A computable function

$$\rho = \rho(\varphi) : \mathbb{N} \to \{0, 1\}$$

is called a variant rejection or cut off if

$$[V(T) - \bigcup_{\rho(n)=1} V(T(\varphi(n)))] \cap \text{Solv } T \neq \phi$$

(which means that we preserve solutions). Here $T(\varphi(n))$ is $T(U)$ from 3.3 b) for $U = \varphi(n)$.

Given a variant rejection ρ for a search (T, φ, F), one can construct a new search $(T, \tilde{\varphi}, F)$ in the following manner. Set

$$\varphi'(n) = \varphi(n) \text{ if } \varphi(n) \notin \bigcup_{\substack{\rho(i)=1 \\ i<n}} V(T(\varphi(i)))$$

$\varphi'(n)$ is not defined otherwise.

Now construct a monotonic numeration $\mu : \mathbb{N} \to \mathbb{N}$ of the points $n \in \mathbb{N}$, for which $\varphi'(n)$ is defined, and suppose moreover that if n is in the image of μ then all $i < n$ are in the image of μ. Then set

$$\tilde{\varphi}(n) = \varphi'(\mu(n))$$

4. Mass problems.

The above notions become somewhat more interesting if one considers them in the case of mass problems.

Suppose we are given a tree T and a family of exhaustions $\mathcal{E} = \{(T, \varphi, F)\}$. It is usual to subject the elements of $\mathcal{E}$ to conditions of coherence. Namely, we require the existence of a computable function

$$\pi : \mathbb{N} \times \mathcal{E} \to V(T)$$

such that the restriction of π to the fiber of $\mathbb{N} \times \mathcal{E}$ over $(T, \varphi, F) \in \mathcal{E}$:

$$\pi : \mathbb{N} \times (T, \varphi, F) \to V(T)$$

coincides with φ (and is therefore subject to conditions a), b) of 3.4). Moreover, we have to assume that our exhaustive search does not depend on the future, that is, if $(T, \varphi, F), (T, \varphi_1, F_1) \in \mathcal{E}$ and $\bigcup_{i \leq n} \varphi(i) = \bigcup_{i \leq n} \varphi_1(i)$ and $F|\bigcup_{i \leq n} \varphi(i) = F_1|\bigcup_{i \leq n} \varphi_1(i)$ then $\pi(n+1, (T, \varphi, F)) = \pi(n+1, (T, \varphi_1, F_1))$.

Now we can also define the variant rejection as a computable function

$$\rho : \mathbb{N} \times \mathcal{E} \to \{0, 1\}$$

such that its restriction to every (T, φ, F) is a variant rejection in the sense of 3.5 and which satisfies the additional conditions stated below.

One is the following. If $(T, \varphi, F) \in \mathcal{E}$, $(T, \varphi_1, F_1) \in \mathcal{E}$ and $\bigcup_{i \leq n} \varphi(i) = \bigcup_{i \leq n} \varphi_1(i)$ and $F|\bigcup_{i \leq n} \varphi(i) = F_1|\bigcup_{i \leq n} \varphi_1(i)$ then $\rho(n+1, (T, \varphi, F)) = \rho(n+1, (T, \varphi_1, F_1))$.

The second one is the requirement that the new family of searches constructed from $\mathcal{E}$ with the help of the family ρ of variant rejections (as in 3.5) forms a family of searches, subject to the condition of independence from the future. Then one can use variant rejections to construct new searches.

5. Some examples of variant rejections.

5.1. Suppose that F is monotonic in the following sense

$F(U_1) > F(U_2)$ implies $F(W_1) > F(W_2)$ for all $W_1 \in f(U_1)$ and all $W_2 \in f(U_2)$.

In this case one can set $\rho(n) = 1$, if there exists $m < n$ such that

$$F(\varphi(m)) > F(\varphi(n)),$$

and set $\rho(n) = 0$ otherwise.

5.2. Suppose a finite group G acts on T in such a way that

$$F(ga) = F(a) \text{ for } a \in V(T)$$

(Here we consider F as a function on T). Suppose we have an algorithm of canonization, which ascribes to every element $t \in V(T)$ a point Canon t on the same orbit $G \cdot t$ of G as t, and such that Canon t = Canon s for $s \in G \cdot t$.

Set

$$\rho(n) = 1 \text{ if } \varphi(n) \neq \text{Canon } \varphi(n)$$

This approach is used in Section R below.

It is useful to find some algorithm Canon, or to construct a search φ such that if $\varphi(n)$ = Canon $\varphi(m)$ then $n \leq m$. Otherwise, the use of the above variant rejection may make the search less effective.

6. Examples of some searches.

6.1. Checkers.

Let a position A on a board be given. Consider the problem of finding a move for black which leads to the best position for white (in some fixed sense) among all positions which are at a distance of 3 successive moves from A.

In this case, V is the set of all positions on the board which can be obtained from the given one in 3 steps with the first step made by black. V is not given explicitly (cf. 2.2) and the rules to construct subsets (checkers moves) are fixed (cf. 2.8). So the tree is constructed uniquely. However, we are free to chose the order in which vertices should be searched (i.e., function φ).

6.2. Strongly regular graphs.

Problem: find all (up to isomorphism) non-empty and non-complete graphs on n vertices, such that every vertex has m neighbors, and two vertices which are (resp. are not) neighbors have the same number d_1 (resp. d_2) of common neighbors.

The set V is the set of all graphs on n vertices. The rules for constructing subsets are rather arbitrary. Let us describe two possible choices.

6.2.1. **Let** V(i) be the set of the graphs on n vertices, such that every one of the first i vertices is incident to m edges, and there are no edges between the vertices n-i+1, ..., n, and any two vertices which are (resp., are not) neighbors have $\leq d_1$ (resp., $\leq d_2$) common neighbors. For a graph $\Gamma \in V(i)$, the associated subset of V consists of all elements of V which coincide with Γ on all edges from the first i vertices. The set $Q_\Gamma = f(\Gamma)$ (cf. 3.3) consists of all graphs from V(i+1) whose first i vertices have the same connections as Γ.

6.2.2. Let V'(i) be the set of all graphs Γ on n vertices, such that there are no edges from the first i vertices to the remaining n-i, and at most m edges from any one of the first i vertices, and two vertices which are (resp., are not) neighbors have $\leq d_1$ (resp., $\leq d_2$) common neighbors. For a graph $\Gamma \in V'(i)$ the associated subset of V consists of all elements of V which coincide with Γ on the first i vertices. The set $Q_\Gamma = f(\Gamma)$ (cf. 3.3) consists of all graphs from V'(i+1) which have the same connections between the first i vertices as Γ.

R. AN ALGORITHM OF GRAPH CANONIZATION.

1. Below we show how the notions and approaches introduced in this volume can be used to describe an algorithm of graph identification. The algorithm of this section is not aimed to be programmed, and therefore it may use "breadth first search" (cf. 8.1). The use of this type of search permits one to apply stabilization of depth 1 or more. The decisive point is the use of Theorem O3.5 to find some orbits of the automorphism group of the graph under consideration.

Another essential feature is the procedure designed to deal with correct graphs (cf. 5.4.2 and 6.2) and direct sums (cf., 5.4.1 and 6.1). Below in sub-section 9 it is explained why this case requires special treatment.

As was mentioned in the introduction, the algorithm of this section is a development of the algorithm of [We 3].

2. Definitions.

2.1. Canonical algorithm

An algorithm $\mathcal{A}$, mapping the set $\mathcal{X}$ of graphs into itself, and defined everywhere on $\mathcal{X}$, is called canonical if

a) for all $X \in \mathcal{X}$ and for all $g \in \mathrm{Sym}\, V(X)$

$$\mathcal{A}(gXg^{-1}) = \mathcal{A}(X)$$

A canonical algorithm $\mathcal{A}$ is called a canonization algorithm if

b) for any X there exists $g \in \mathrm{Sym}\, V(X)$ such that

$$\mathcal{A}(X) = gXg^{-1}$$

2.1.1. Remark. If one has an algorithm $\mathcal{A}$ of graph canonization, then it gives rise to an algorithm of graph identification, say B. Namely B consists of the application of $\mathcal{A}$ to both graphs, and then in the comparison of the results.

2.2. Semi-invariant algorithm

An algorithm $\mathcal{A}$ mapping $\mathcal{X}$ into itself is called <u>semi-invariant</u> if for any $X \in \mathcal{X}$ and for any $g \in \text{Sym } V(X)$ there exists $h \in \text{Aut } X$ such that

$$\mathcal{A}(gXg^{-1}) = gh\mathcal{A}(X)h^{-1}g^{-1}$$

(Recall: $\text{Aut } X = \{h \in \text{Sym } V(X) \; : hXh^{-1} = X\}$.)

The semi-invariant algorithms lie somewhere between invariant algorithms and canonization algorithms. If h is always unity, then $\mathcal{A}$ is invariant.

3. Below we describe how one can construct canonization algorithms if some special kind of semi-invariant algorithms is given.

3.1. Let $\mathcal{A}$ be a semi-invariant algorithm which places a split graph $\mathcal{A}(X)$, whose variables are linearly ordered into correspondence with a graph X. Let $m = m(X)$ be the permutation such that the diagonal entries of $m\mathcal{A}(X)m^{-1}$ are positioned in decreasing order, that is, if $m\mathcal{A}(X)m^{-1} = (y_{ij})$ then $y_{ii} > y_{i-1,i-1}$. Put

$$\text{Canon}_{\mathcal{A}}(X) = mXm^{-1}.$$

3.2. <u>Assertion.</u> The map $X \to \text{Canon}_{\mathcal{A}}(X)$ is a canonization algorithm.

<u>Proof.</u> The validity of 2.1b) is ensured by construction. Now, if $Y = gXg^{-1}$ then $\mathcal{A}(Y) = gh\mathcal{A}(X)h^{-1}g^{-1}$ for some $h \in \text{Aut } X$, since $\mathcal{A}$ is semi-invariant. Further, since both $\mathcal{A}(X)$ and $gh\mathcal{A}(X)h^{-1}g^{-1}$ are split, the substitutions $m = m(X)$ and $m_1 = m(gXg^{-1})$ are uniquely defined. Hence the diagonals of $m\mathcal{A}(X)m^{-1}$ and $m_1gh\mathcal{A}(X)\cdot h^{-1}g^{-1}m_1^{-1}$ coincide. Therefore $m = m_1gh$. Now we have

$$\text{Canon}_{\mathcal{A}}(gXg^{-1}) = mh^{-1}g^{-1}(gXg^{-1})ghm^{-1} = mh^{-1}Xhm^{-1} = mXm^{-1} = \text{Canon}_{\mathcal{A}}(X)$$

as desired. Note that the third equality used the condition $h \in \text{Aut} X$.

4. Our canonization algorithm consists of several parts.

The following two parts are most important:

4.1. <u>Splitting algorithm.</u>

It is an exhaustive algorithm. The vertices of the associated tree (cf. Q3.3) correspond to the stationary graphs. The set Q_X (cf. Q3.3) is either the set $\{\lambda_m(X)\}_{m \in V(X_{ii})}$ or the disassemblage of a correct graph X (cf. J6.7). In addition, at a vertex of our tree the disassemblage of direct sums is possible. End points of our tree correspond to split graphs with linearly ordered entries.

4.2. Elevation algorithm.

This part works on the results of the job done by the preceding algorithm.

Using either the stabilization of depth 1 or O3.5, the elevation algorithm gradually, step by step, decreases the depth of the tree of 4.1. At every moment, however, endpoints correspond to split graphs. At the end of its work the elevation algorithm delivers a split graph. Then a canonical form is constructed according to 3.1. Note, that the elevation algorithm is invariant only in the case Aut X = 1; in the general case it is semi-invariant (cf. 6.1, 6.2, 6.4 below).

5. Splitting algorithm (denoted Split).

5.1. Let T be a directed tree, T(k) be its k-th level. If $v \in T(k)$ then let Q_v denote those vertices of T(k+1) which are connected with v; let P_v denote the vertex from T(k-1) such that (P_v, v) is an edge of T. Further, let T_v denote the tree hanging at v.

5.2. In our case T is the tree of the exhaustive algorithm Split. To every vertex $v \in T$ there corresponds a stationary graph denoted by X(v).

5.2. If $v \in T$ then X(v) is checked for validity of the following conditions:

5.3.1. X(v) decomposes into a direct sum (cf. G2);

5.3.2. X(v) is correct (cf. J6.6).

5.4. Q_v, $v \in T$, is defined in the following manner:

5.4.1. If X(v) decomposes into a direct sum, $X(v) = \widetilde{\bigoplus}_{t(v) \geq i \geq 1} Y_i(v)$, $Y_{i+1}(v) > Y_i(v)$, then $Q_v = \{Y_1(v), \ldots, Y_{t(v)}(v)\}$.

5.4.2. If 5.4.1 is not applicable but X(v) is correct, then $Q_v = F(X(v))$, cf. J6.7.

5.4.3. If 5.4.1 and 5.4.2 are not applicable, $X = (X_{ij})$ and (composition of $X_{tt}) = \max_{i:|X_{ii}|>1}$ (composition of X_{ii}) then $Q_v = \{\lambda_m(X)\}_{m \in V(X_{tt})}$.

5.5. Remarks.

5.5.1. Note that for $v \in T$ the tree $T(v)$ corresponds to the algorithm Split, applied to $X(v)$.

5.5.2. Also note that in the case 5.4.1 there is no branching of the exhaustive search at the point under consideration since $X = X_1 \tilde{\oplus} X_2$ implies (as we shall see below) that Canon X = Canon $X_1 \tilde{\oplus}$ Canon X_2.

6. Elevation algorithm (denoted Lift).

6.1. Assemblage of a direct sum.

Let $\mathcal{A}$ be an algorithm on graphs. Let $X = \tilde{\oplus}_{m \geq i \geq 1} X_i$. Renumber the variables of the graphs $\mathcal{A}(X_i)$ according to the lexicographical order of (X_i, x_j) for $x_j \in \mathcal{A}(X_i)$. Denote the result by $\mathcal{A}'(X_i)$.

Denote by Assembly (X) the graph, obtained in the following manner:

In the matrix X, replace any entry of X_i by the corresponding entry of $\mathcal{A}(X_i)$. Then $\text{Assembly}_{\mathcal{A}}(X)$ is the stabilization of this latter graph.

Note that if $\mathcal{A}(X_i)$ is split for all i, then $\text{Assembly}_{\mathcal{A}}(X)$ is also split. If $\mathcal{A}$ is semi-invariant then $\text{Assembly}_{\mathcal{A}}$ is also semi-invariant. If $\mathcal{A}$ is semi-invariant and $\mathcal{A}(X_i) = R$ for all i, then

$$\text{Canon}_{\text{Assembly}_{\mathcal{A}}}(X) = \text{Assembly}_{\text{Canon}_{\mathcal{A}}}(X).$$

6.2. Assemblage of correct cellular algebras.

Let X be a correct stationary graph which can not be decomposed into a direct sum. Let $F(X) = \{X_i\}$ be its disassemblage. For any algorithm on graphs one can define their assemblage in the same manner as in 6.1. However we shall define it only for semi-invariant algorithms $\mathcal{A}$ which bring a split graph $\mathcal{A}(X)$ in correspondence with X. In this case let us order the graphs X_i according to the (lexicographical) order of the graphs Canon (X). By J6.8 isomorphic graphs (that is, those for whom $\mathrm{Canon}_{\mathcal{A}}(X_i) = \mathrm{Canon}_{\mathcal{A}}(X_j)$) define some subgroup in Aut X which permutes isomorphic graphs as an appropriate group Sym. Therefore, if we define an arbitrary order within the isomorphism class we obtain a semi-invariant algorithm.

Explicitly, for $\mathcal{A}(X_i) = (x^i_{pq})$ construct $\overline{X}_i = (\overline{x}^i_{pq})$ in the following manner:

$$\overline{x}^i_{pq} > \overline{x}^i_{rt} \quad \text{if} \quad x^i_{pq} > x^i_{rt}$$

$$x^i_{pq} > x^j_{rt} \quad \text{if} \quad \mathrm{Canon}_{\mathcal{A}}(X_i) > \mathrm{Canon}_{\mathcal{A}}(X_j) \quad \text{or if}$$

$$\mathrm{Canon}_{\mathcal{A}}(X_i) = \mathrm{Canon}_{\mathcal{A}}(X_j) \quad \text{but} \quad i > j.$$

Again let us note that in the latter case the definition is semi-invariant by J6.8. If Aut X = 1, then the latter case does not occur and the definition is invariant.

Now denote by $\overline{X}$ the matrix obtained by substitution in X of the entries of $\overline{X}_i$ for the corresponding entries of X_i. Set

$$\mathrm{Assembly}_{\mathcal{A}}(X) = \mathrm{Stab}\,\overline{X}.$$

It is a semi-invariant algorithm.

6.3. <u>Elevation in the case</u> $Q_v = \{\lambda_i(X)\}$.

Let $\mathcal{A}$ be a semi-invariant algorithm which maps every graph into a split graph with linearly ordered entries. Let I be the set of all i such that for all j one has $\mathrm{Canon}_{\mathcal{A}}(\lambda_i(X)) \geq \mathrm{Canon}_{\mathcal{A}}(\lambda_j(X))$. In particular, all graphs $\lambda_i(X)$, $i \in I$, are isomorphic. Set

$$m = \max_{i \in I} i \quad \text{and} \quad \mathrm{Change}_{\mathcal{A}}(X) = \mathcal{A}(\lambda_m(X)).$$

This operation is semi-invariant by O3.5. If Aut X = 1 then $|I| = 1$, and our operation is invariant.

6.4. Define inductively the <u>elevation algorithm</u> Lift.

Let $\widetilde{T}$ be a subtree of T and denote the graph corresponding to the vertex v of $\tilde{T}$ by $\tilde{X}(v)$. Suppose that $\tilde{X}(v) = X(v)$, and $Q_v(\tilde{T}) = Q_v(T)$ if $v \notin \mathrm{End}\,\tilde{T}$. $\tilde{X}(v)$ is split if $v \in \mathrm{End}\,\tilde{T}$.

The tree T constructed for algorithm Split satisfies these conditions.

Let us construct from $\tilde{T}$ a new tree $\tilde{\tilde{T}}$ in the following manner. Let $v \notin \mathrm{End}\,\tilde{T}$, but $Q_v(\tilde{T}) \subset \mathrm{End}\,\tilde{T}$. Then substitute for $\tilde{X}(v)$ a matrix $\tilde{\tilde{X}}(v)$ obtained from $\tilde{X}(v)$ by the rules 6.1, 6.2, 6.3, if Q_v was obtained from v by the rules 5.4.1, 5.4.2, 5.4.3 respectively. If $v \notin \mathrm{End}\,T$ and $Q_v \not\subset \mathrm{End}\,T$, put $\tilde{\tilde{X}}(v) = \tilde{X}(v)$. Also set $V(\tilde{\tilde{T}}) = V(\tilde{T}) - \mathrm{End}\,\tilde{T}$. Let us denote the pair: the tree $\tilde{\tilde{T}}$ and the map $v \to \tilde{\tilde{X}}(v)$ by Lift $(\tilde{T}, \tilde{X}(v))$. Algorithm Lift is evidently semi-invariant (cf. 6.1, 6.2, 6.3). Since Lift $(\tilde{T}, \tilde{X}(v))$ satisfies the conditions on $\tilde{T}$, we can apply Lift recursively.

7. Canonization

7.1. Let A be an arbitrary matrix. Construct (cf. M2.1) X(A).

7.2. Let X = Stab X(A). Construct for X the splitting algorithm (cf. 5), its tree T, and the correspondence $v \to X(v)$.

7.3. Let N be the maximal length of the paths in T. Then $|V((\mathrm{Lift})^N(T))| = 1$. Let $v_0 = V((\mathrm{Lift})^N(T))$, $\hat{X} = (\mathrm{Lift})^N(X(v_0))$. The correspondence $X \to \hat{X}$ is a semi-invariant algorithm. Denote it by $\mathcal{A}$.

7.4. Construct $\mathrm{Canon}_{\mathcal{A}}(A)$ (cf. 3.1).

8. Variants

8.1. Simultaneous descent (Breadth-first search).

It is possible to construct for the tree T of the splitting algorithm the entire following level T(k+1). Then one can perform the simultaneous stabilization of all graphs at that level and compare the results. Moreover one can in this case perform the stabilization of depth (k+1) (cf. O6.2). In such an approach a picture would be more homogeneous and natural (but quite impractical, cf. Q.2.5).

8.2. Successive descent (depth first search).

It is possible to first descend up to the end of the most left branch of the tree T, and then move to the right. Such an approach allows one to save memory (cf. Q2.5). Besides, with luck, one can find rather early automorphisms, and then use them in the same manner as in S3.3.

9. Some explanations.

When one is trying to handle the graph isomorphism problem, he first uses the ideas introduced in the stabilization algorithm. Then he applies the same ideas for graphs with one, two and so forth fixed vertices (i.e., an analogue of 5.4.3 above). Clearly this leads to a solution of the problem, but sometimes it can take too long to get to this solution. The first evident obstacles are direct sums and correct graphs.

9.1. If $X = Y \tilde{\oplus} Z$ then fixation of the vertices of Y does not affect Z, and vice versa. So the depth of the corresponding tree of exhaustion is the sum of depth of the trees for Y and Z. Then the "time" required for such an approach is the product of the "times" required for Y and Z.

However in the approach we used, we are dealing with Y and Z separately, and so the "time" is only the sum of the "times" for Y and Z.

9.2. Analogously, if X is a correct graph and $\{X_i\}$ its disassemblage, then our approach again requires only the sum of the "times" for each X_i, but a "straightforward" approach requires a product of "times". The simplest case is when X is a simplex, S_m. Then the straightforward approach requires $m!$ steps. Our approach requires one step.

However in the case of a simplex one can use automorphisms (as described in S3.5) and get the result in less than $m!$ steps. However, some correct graphs have no automorphisms permuting their parts and then once again one gets a factorial.

9.3. Remark. Since some $\lambda_i(X)$ could be direct sums, or correct graphs, even if X is neither, 5.4.1 and 5.4.3 could be used repeatedly by our algorithm.

5. A PRACTICAL ALGORITHM OF GRAPH CANONIZATION.

The algorithm for the construction of strongly regular graphs, which is described in the next Section, constructs many (thousands of) graphs. Therefore a program was written which canonized graphs constructed by the algorithm of the next Section. We describe below the ideas on which this program was based (we follow the exposition given in [Ar 1]). Note that strongly regular graphs are rather difficult to handle, because they have a high degree of symmetry. On the other hand the algorithm of the preceeding Section is too bulky for practical purposes. An interesting feature of the algorithm of this Section is the procedure designed to construct and to use automorphisms of the graph, cf. 3. below.

1. For the $n \times n$ (0,1)-matrices $A = (a_{ij})$ and $B = (b_{ij})$ with zero diagonal let us write

$$A > B$$

if the n^2-dimensional vector $(a_{11}, a_{12}, \dots, a_{nn})$ is greater (in lexicographical order) than $(b_{11}, b_{12}, \dots, b_{nn})$. Let us then say that

$$\tilde{A} = \max_{g \in \mathrm{Sym}(n)} gAg^{-1}$$

is the <u>maximal form</u> of A.

We shall construct an algorithm $\mathcal{A}$ (mapping the set of symmetric (0,1)-matrices with zero diagonal into itself) such that

$$\mathcal{A}(A) = \tilde{A} .$$

Such an $\mathcal{A}$ will be a canonization algorithm (in the sense of the preceeding Section).

2. For an ordered subset $V(k) = (i_1, \dots, i_k)$ of distinct elements of $[1, n]$ and for a matrix A of the same type as in 1 above, put (Min stands for "Minor")

$$\mathrm{Min}_{V(k)}(A) = (a_{st})_{(s,t) \in V(k) \times V(k)} .$$

Let us say also that $V(n) = (i_1, \dots, i_n)$ is a <u>monotonic sequence</u> for A if for every $\bar{V}(k) = (i_1, \dots, i_k)$ one has

$$\mathrm{Min}_{\bar{V}(k+1)}(A) = \max_{V(k+1) \mid V(k+1) \supset \bar{V}(k)} \mathrm{Min}_{V(k+1)}(A)$$

The relation of the notions of a monotonic sequence and a maximal form is explained by

2.1. <u>Proposition</u>. Let A be a symmetric $(n \times n)$ $(0,1)$-matrix with zero diagonal. If A is a maximal form, then the sequence $(1, \dots, n)$ is monotonic for A.

<u>Proof</u>. Suppose that the assertion is false, let $k \in [1, n]$ be the smallest number such that for $\bar{V}(k+1) = \{1, \dots, k+1\}$ one can find $s > k+1$ such that for $V(k+1) = (1, \dots, k, s)$ one has

$$\mathrm{Min}_{\bar{V}(k+1)}(A) < \mathrm{Min}_{V(k+1)}(A)$$

Let m be the first number s with these properties. Then we have for some $t < k$ (recall, that A has zero diagonal) that

$$a_{im} = a_{i,k+1} \text{ for } i = 1, 2, \dots, t \tag{1}$$

but

$$a_{t+1,m} > a_{t+1,k+1} \tag{2}$$

Let $g \in \mathrm{Sym}(n)$ be the transposition of m and $k+1$. Then the first t rows of gAg^{-1} coincide with the first t rows of A (it follows from (1)) but the $(t+1)$-th row of gAg^{-1} is greater than the $(t+1)$-th row of A (this follows from (2)). Therefore A is not a maximal form. This is a contradiction.

2.2. Proposition 2.1 shows that to find a substitution $g \in \mathrm{Sym}(n)$ such that gAg^{-1} is the maximal form of A, one need not consider all $g \in \mathrm{Sym}(n)$, but only those for which the subset $V(n) = g(1, \dots, n)$ is monotonic for A.

This can be done rather easily since we have to consider only those $V(k)$ which are monotonic for $\mathrm{Min}_{V(k)}(A)$.

More precisely, let A be a $n \times n$ symmetric $(0,1)$-matrix with zero diagonal and suppose that $V(k)$ is monotonic for $\mathrm{Min}_{V(k)}(A)$. Let us call $V(k+1) = (i_1, \ldots, i_{k+1})$ an extension of $V(k)$ if

a) $V(k) = (i_1, \ldots, i_k)$

b) $V(k+1)$ is monotonic for $\mathrm{Min}_{V(k+1)}(A)$.

Therefore we have to take all extensions of $V(0) = \phi$, then all extensions of these extensions,and so forth,until we will get the end points, which are monotonic sets for A consisting of n elements. Every such set $(i_1, \ldots, i_n)$ determines the substitution $g = \begin{pmatrix} 1 & 2 & & n \\ i_1 & i_2 & \cdots & i_n \end{pmatrix}$ and $(1, 2, \ldots, n)$ is a monotonic sequence for gAg^{-1}. Using Proposition 2.2 one has only to chose the greatest matrix among the matrices gAg^{-1} described above.

The algorithm for the construction of extensions of a given set will be described in 2.4.

2.3. Let us now show how the description of this exhaustive search is interpreted in terms of Section Q.

Let $T(k)$ be the set of $V(k)$ such that $V(k)$ is monotonic for $\mathrm{Min}_{V(k)}(A)$. Set $T = \bigcup T(k)$. There is an edge from $t \in T(k)$ to $s \in T(k)$ if $s = V(k+1)$ is an extension of $t = V(k)$. Since for monotonic $V(k+1) = (i_1, \ldots, i_{k+1})$, the sequence $V(k) = (i_1, \ldots, i_k)$ is also monotonic, and the tree T is connected. It is clear that every monotonic sequence $V(k)$ (for $\mathrm{Min}_{V(k)}(A)$) has at least one extension. Therefore every monotonic sequence for A is represented by an end point of T, and every end point represents a monotonic sequence for A.

2.4. Let us now describe the exhaustion function of our search, that is the algorithm for constructing extensions of sets. This part is repeated many times and therefore has to be as effective as possible.

Let $V(k) = (i_1, \dots, i_k)$ be a monotonic sequence for $\mathrm{Min}_{V(k)}(A)$. If $V(k+1) = (i_1, \dots, i_k, r)$ is an extension of $V(k)$, then it follows from the definition of the monotonic sequence that

$$(a_{i_1, r}, \dots, a_{i_k, r}) = \max_{t \in [1, n]-V(k)} (a_{i_1, t}, \dots, a_{i_k, t})$$

Let $R = R(V(k))$ be the set of all r which satisfy this condition (then $(V(k), r)$, $r \in R$, are all extensions of $V(k)$).

To describe R we use the sets $W_i = \{j \in [1, n] - V(k) \mid a_{ij} = 1\}$. Put $R_0 = \{1, 2, \dots, n\} - V(k)$ and

$$R_s = \begin{cases} R_{s-1} & \text{if } R_{s-1} \cap W_{i_s} = \phi \\ R_{s-1} \cap W_{i_s} & \text{if } R_{s-1} \cap W_{i_s} \neq \phi \end{cases}$$

<u>Proposition</u>. R_s is the set of $j \in [1, n] - V(k)$ such that

$$a_{i_1, j}, \dots, a_{i_s, j} = \max_{t \in [1, n]-V(k)} (a_{i_1, t}, \dots, a_{i_s, t})$$

The proof is straightforward (cf. also [Arl]).

Using this proposition one can find R by only taking intersections of computer words.

3. For some graphs the above procedure is ineffective. The graphs whose automorphism groups are large will have very large tree T. Another case is the case of correct graphs. We shall show below how to deal with a large automorphism group (cf., T5.2).

Here we have two problems. The first one is how to find automorphisms, and the second one is how to use them. The following two assertions answer these questions. They are evident.

3.1. <u>Proposition</u>. Suppose that $\mathrm{Min}_{V(n)}(A) = \mathrm{Min}_{\widetilde{V}(n)}(A)$, $V(n) = (i_1, \dots, i_n)$

$\tilde{V}(n) = (j_1, \ldots, j_n)$. Then the permutation

$$g = \begin{pmatrix} i_1, \ldots, i_n \\ j_1, \ldots, j_n \end{pmatrix}$$

is an automorphism of A.

3.2. Proposition. Let $g \in \mathrm{Aut}\, A$ and $V(k) = (i_1, \ldots, i_k)$ be such that $V(k)$ is monotonic for $\mathrm{Min}_{V(k)}(A)$ and $gi_j = i_j$ for $j = 1, \ldots, k$. Let $T(U)$, $U \subset [1, n]$, be the subtree of the tree T from 2.3, consisting of $\tilde{U} \subset [1, n]$ such that $\tilde{U} = \{U, j_1, \ldots, j_t\}$. Then for every $j \in [1, n] - V(k)$ and every $V(n) \in T(\{V(k), j\})$ one has $\mathrm{Min}_{V(n)}(A) = \mathrm{Min}_{gV(n)}(A)$ and $gV(n) \in T(\{V(k), gj\})$.

3.3. To use the preceeding assertions, we use the "depth first search" over the tree T described in 2.3. All sequences $V(n) = (i_1, \ldots, i_n)$, which are monotonic for A, are stored together with the matrices $gAg^{-1} = \mathrm{Min}_{V(n)}(A)$, where $g = \begin{pmatrix} 1 & \ldots & n \\ i_1 & & i_n \end{pmatrix}$. When a new monotonic sequence $V(n)$ is constructed, we compare the corresponding matrix gAg^{-1} with the already stored matrices. If it coincides with one of them, then we get (by Proposition 3.1) an automorphism.

For every $V(k)$ belonging to a sequence $V(0) = \phi \subset V(1) \subset \ldots$ of extensions, and for every $g \in \mathrm{Aut}\, A$ found by the above method and such that $g|V(k) = 1$, let us store the orbits of g in the set of extensions of $V(k)$. When a new such g is found, the intersecting orbits are joined.

Now Proposition 3.2 says that in the search we can take only one representative of the extensions of $V(k)$ (compare T5.2).

3.4. The method described in 3.3 is heuristic (cf. Q2.11). It is useful when the group Aut A is large. If it is small (e.g., $\mathrm{Aut}\, A = \{1\}$) then all our efforts (and storage space) will be useless.

3.5. An example of a situation where 3.3 essentially reduces our search, is the case $A = \tilde{I}_n$ (the complete graph). In this case the heuristic of 3.3 requires the construction of only n end points of T (but the method of 2.3

requires the construction of $n!$ end points of T since every sequence $g(1, \ldots, n)$, $g \in \mathrm{Sym}(n)$ is monotonic for A).

3.6. However in the case of correct graphs (cf., J6) the heuristic of 3.3 may fail for reasons described in 3.4, and the size of the search would be of order $(\frac{n}{a})!$, which is still very large for graphs of G4.7, for example.

4. The algorithm described in this Section (with heuristic 3.3) was used to canonize the graphs constructed by the algorithm of the next Section. It also found the orbits of the automorphism group.

For the graph # 7 from the 26-family, the program based on this algorithm constructed 40 end points of T and for # 9 from the 26-family it constructed 756 end points of T. Note that Aut A is trivial in the second case.

T. AN ALGORITHM OF CONSTRUCTION OF STRONGLY REGULAR GRAPHS.

A strongly regular graph with parameters $n, n_1, a^1_{11}, a^2_{11}$ is a graph with n vertices, such that

a) any vertex is incident to n_1, $0 < n_1 < n-1$, vertices,

b) any pair of incident vertices is simultaneously incident to a^1_{11} different vertices,

c) any pair of non-incident vertices is simultaneously incident to a^2_{11} different vertices.

Clearly the adjacency matrix $A = (a_{ij})$ of a strongly regular graph is a basic element of a three-dimensional cell (cf. K20). So it is a symmetric $n \times n$-matrix with zero diagonal and with elements 0 and 1, and it satisfies the following condition:

If $s_i = (a_{i1}, \ldots, a_{in})$ is the i-th row of A and if $|s|$ denotes the number of ones in (0,1)-vector s, then

$$|s_i| = n_1 \quad \text{for all } i$$
$$|s_i \cap s_j| = \begin{cases} a^1_{11} & \text{if } a_{ij} = 1 \\ a^2_{11} & \text{if } a_{ij} = 0 \end{cases} \qquad (*)$$

Below we describe the algorithm which constructs for a given set $n, n_1, a^1_{11}, a^2_{11}$ of parameters a set of strongly regular graphs with these parameters such that any strongly regular graph with these parameters is isomorphic to at least one constructed graph. Interesting features of our algorithm are the use of partial canonization, cf. 2.3, 2.5, 2.6, 2.8, and two forced variants (cf. 2.4, 3.4).

1. To describe this algorithm we have to introduce some notions. Let us fix n. Let $B = (b_{ij})$ be a $n \times n$ (0,1)-matrix with zero diagonal. Let $D = \{i_1, \ldots, i_t\}$ be a subset of $[1, n]$.

Two numbers $r, q \in [1,n] \setminus D$ are said to be D-equivalent if

$$b_{iq} = b_{ir} \quad \text{for all } i \in D$$

(clearly this is an equivalence relation).

1.1. Definition. Set $D_0 = D$ and let D_i, $i = 1, 2, \ldots, m$, be classes of D-equivalence numbered such that (here inf stands for infimum)

$$i > j \Leftrightarrow \inf D_i > \inf D_j$$

The D_i's are called D-sets.

1.2. Let B, D, D_i be as above. Let $s_q = s_q(B)$ be the q-th row of B. For $q \notin D$ and $j \in [1, m]$ set

$$x_{qj} = x_{qj}(B) = \sum_{d \in D_j} b_{qd}.$$

(This is the number of ones in s_q which occupy the positions (q, s), $s \in D_j$.)

1.3. Proposition. Let $p = \inf D_i$.

a) For every $q \in D_i$ there exists $g \in \mathrm{Sym}(D_i)$ such that

$$x_{pj}(g^{-1}Bg) = x_{qj}(B) \quad \text{for all } j \in [1,m].$$

b) There exists an $h \in \prod_{i \geq 1} \mathrm{Sym}(D_i - p)$ such that for $h^{-1}Bh = (c_{\alpha\beta})$ the following holds: $c_{ps} = 1$, $s \in D_k$, implies $c_{pj} = 1$ for all $j < s$, $j \neq p$, $j \in D_k$.

These assertions are evident, and show that, when t rows $(i_1, \ldots, i_t)$ of B are fixed, we still have some freedom to move the remaining rows. They also show how to use this freedom. These assertions, and also their corollary, are used in 2.3, 2.5, 2.6.

An easy corollary of 1.3a) is

1.4. Corollary. Under assumptions and notations of 1.3 there exists $g \in \mathrm{Sym}(D_i)$ such that the vector

$$(x_{p1}(gBg^{-1}), \ldots, x_{pt}(gBg^{-1}))$$

is lexicographically greater than or equal to any vector

$$(x_{q,1}(gBg^{-1}),\dots,x_{q,t}(gBg^{-1})), \qquad q \in D_i .$$

1.5. Now note that the number b_{rj}, $r \in D$, $j \in D_i$, depends only on r and i. Let us denote it by $b_i(r)$.

Secondly, if we set $s_{k,D} = (b_{k,i_1},\dots,b_{k,i_t})$, (recall that $D = (i_1,\dots,i_t)$), then the number

$$|s_{r,D} \cap s_{j,D}|, \quad r \in D,$$

does not depend on $j \in D_i$. Let us denote it by $c_i(r)$.

Also let c_i be the common value of $|s_{j,D}|$, $j \in D_i$.

1.6. Proposition. If B is an adjacency matrix of a strongly regular graph, then the numbers $x_{qj}(B)$, $q \in D_i$, satisfy the following equations

$$\sum_{j>0} x_{qj} = n_1 - c_i$$

$$\sum_{j>0} b_j(r) x_{qj} = \begin{cases} a^1_{11} - c_i(r) & \text{if } b_i(r) = 1 \\ \\ a^2_{11} - c_i(r) & \text{if } b_i(r) = 0 \end{cases}$$

(These relations are direct consequences of (*)).

2. Now we are able to describe the work of our algorithm at a fixed vertex (i.e., to describe its function of an exhaustion).

There are two somewhat different procedures depending on the situation. In all cases at the level t, the data inherited from the level $(t-1)$ contain a subset $D(t) = \{i_1,\dots,i_t\}$ of $[1,n]$, a $n \times n$ $(0,1)$-matrix B_t and also some additional information to be described later ($S_{d,t}$, a list of positions fixed (cf. 2.4) at the level $t-1$).

2.1. For this pair $D(t), B_t$ construct $D(t)$-sets $D_1,\dots,D_m$ numbered as in 1.1. If

$|D_i| = 1$ for all $i = 1, \ldots, m$, the second procedure is applied. It is described in 3.

2.2. For every $i = 1, 2, \ldots, m$, find all solutions of the system

$$\sum_{j \geq 1} x_j = n_1 - c_i$$

$$\sum_{j \geq 1} b_j(r) x_j = \begin{cases} a_{11}^1 - c_i(r) & \text{if } b_i(r) = 1 \\ \\ a_{11}^2 - c_i(r) & \text{if } b_i(r) = 0 . \end{cases}$$

The solutions are vectors of length m. Let us denote the set of these solutions by S_i. The elements of S_i are ordered with respect to dictionary order.

The search for solutions of the above systems is done by the evident exhaustive search. If the procedure "Fixation" (cf. 2.4) has already determined (that is on preceeding levels) the values of some x_j, those values are not computed anew but substituted in the above systems. ("Fixation" reduces our search, but does not involve an exhaustive search. Therefore, "Fixation" is a forced variant).

If for at least one i the above system has no solutions, return to the level $t - 1$ and apply 2.6.

2.3. Now fix in turn all $s < t$ and all $i \in [1, m]$. Let $D_1', \ldots, D_a'$ be D(s)-sets constructed for B. Clearly D_i is contained in some D_j'. If none of $i_{s+1}, \ldots, i_t$ belongs to D_j', pass to 2.4. Otherwise take the largest among $i_{s+1}, \ldots, i_t$ which belongs to D_j', and call it r. For this r, compute the solution of the system of 2.2 which is realized by r-th row of B_t. Let it be the vector u (of length a).

For each solution $v \in S_i$, compute the solution of the system 2.2 (considered for s) to which it corresponds. Denote it by $\bar{v}$ (the computation is easily performed for a given matrix B).

Delete from S_i, those v for which $\bar{v} > u$. If the resulting S_i is empty, return to the level $t - 1$ (i.e., put $t := t - 1$ and apply 2.6).

2.4. "Fixation" (This is an example of a forced variant, cf., Q 2.10). If for some $i, j \in [1, m]$ the j-th component of all solutions from S_i is 0, then we can fill all positions $(p, q) \in D_i \times D_j$ of B by zeros, and we do this.

Analogously, if for some $i, j \in [1, m]$, $i \neq j$ (resp., for some $j \in [1, m]$) the j-th component of all vectors from S_i (resp., S_j) is $|D_i|$ (resp., $|D_j| - 1$), we can fill all positions of $(p, q) \in D_j \times D_i$ (resp., $(p, q) \in D_j \times D_j$, $p \neq q$) by ones, and we do this.

2.4.1. Note, that the group $\mathrm{Sym}\, D_1 \times \ldots \times \mathrm{Sym}\, D_m$ commutes with these fixed pieces.

2.5. Now take the smallest $d \in [1, m]$ such that $|S_d| \leq |S_i|$ for all $i \in [1, m]$ and put $i_{t+1} := \inf D_d$, $S_{d,t} := S_d$.

2.6. Take the largest vector, say v, from $S_{d,t}$ and put $S_{d,t} := S_{d,t} - v$, $D(t+1) = \{i_1, \ldots, i_{t+1}\}$. Place the elements in the i_{t+1}-th row of B_t (they will lie outside D) in the following manner. Let $D_i = \{j_{i,1}, \ldots, j_{i,|D_i|}\}$ (listed in increasing order), $v = (v_1, \ldots, v_t)$. If $i \neq d$, put

$$b_{i_{t+1}, j} = \begin{cases} 1 & \text{for } j = j_{i,1}, \ldots, j_{i,v_i} \\ 0 & \text{for } j = j_{i,v_i+1}, \ldots, j_{i,|D_i|} \end{cases}$$

If $i = d$ put (recall $j_{d,1} = i_{t+1}$)

$$b_{i_{t+1}, j} = \begin{cases} 1 & \text{for } j = j_{d,2}, \ldots, j_{d,v_d+1} \\ 0 & \text{for } j = i_{t+1}, j_{d,v_d+2}, \ldots, j_{d,|D_d|} \end{cases}$$

Then insert the corresponding i_{t+1}-th column (so that B is symmetric).

Also insert the entries in all positions which were determined in 2.4 and store the information about these positions. Call the resulting matrix B_{t+1}.

2.7. If $t + 1 = n$, print B_{t+1} (which is the matrix of a strongly regular graph), set $t := t - 1$ and apply 2.3. If $t + 1 < n$, go to the level $t + 1$, i.e., set $t := t + 1$ and apply 2.1.

2.8. Remark. We have used several search reductions (in 2.3, 2.5, 2.6). Using them we construct a smaller number of graphs. However, at least one graph in each isomorphism class will be constructed. This is guaranteed by 1.4 (for 2.3), 1.3a) (for 2.5), 1.3b) (for 2.6). (cf. also Q5.2 where this situation is described in a more detached way).

3. "Break-down". Now suppose that in the situation of 2.1 one has $|D_i| = 1$ for $i = 1, \ldots, m$ (then $m = n - |D|$). Let t_0 be the first level at which it occurred. In this case the solutions of 2.2 are (0.1)-vectors, and the set of the solutions for level $t + 1$ is easily obtained from the set of the solutions for level t. Indeed, the number of variables decreases and the number of equations increases. Therefore, it is worth to store the set of solutions. The list of solutions is organized as follows:

All solutions (at level t_0)

admissible at the level t	forbidden at the level t	forbidden at the level t-1	forbidden at the level t-2	....

3.1. If $t = t_0$, do the same as in 2.2 and 2.3. If $t > t_0$, move all solutions which were admissible at the level $t - 1$ but contradict the i_t-th row, to the list: forbidden at the level t.

3.2. If the list of admissible solutions at the level t is empty, return to the level $t - 1$.

3.3. The same as 2.4.

3.4. For each solution from the list (if $t > t_0$ use the list: admissible at the level t-1) check whether it contradicts the fixations made in 3.3 (i.e., we check to some extent the list of solutions for compatibility). If there are no contradictions pass to 3.5.

If there are contradictions move contradicting solutions to the list of solutions forbidden at the level t and apply again 3.2. (In this way one again gets a "forced variant".)

3.5. If the list of admissible solutions at the level t is empty and $t > t_0$, set $t := t - 1$ and pass to 3.6.

If the list of admissible solutions at the level t is empty and $t = t_0$, set $t := t - 1$ and return to 2.6.

3.6. If the list of admissible solutions at the level t is not empty, take for the i_{t+1} the least number $i_{t+1} \in D$ such that the number of solutions for the corresponding row is minimal; take from the list of admissible at the level t solutions the largest one corresponding to i_{t+1}-th row; move it to the list of forbidden at level $t + 1$, and insert the corresponding column and row in B_t. Call the resulting matrix B_{t+1}, set $t := t + 1$ and return to 3.1.

4. Let us now describe the tree of our exhaustive search. The vertices of level t are pairs $(D(t), B_t)$ consisting (cf. 2) of a subset $D(t) = \{i_1, \dots, i_t\}$ of $[1, n]$ and a $(n \times n)$ (0.1)-matrix with zero diagonal whose rows with numbers $i_1, \dots, i_t$ satisfy relations (*). (But matrix B_t can contain ones outside the rows and columns with the numbers $i_1, \dots, i_t$.) There is an edge from $(D(t), B_t)$ to $(D(t+1), B_{t+1})$ if the latter pair is constructed from the former one with the help of the rules described in 2.5, 2.6 and 3.6.

The search is the "depth first search".

5. Forced variants are "Fixation" (cf., 2.4) and "Fixation-Deletion" (cf., 3.4). They are helpful for constructing part of the matrix B without branching.

6. The choice of i_{t+1} (cf. 2.5) is heuristic. We do not know and did not know whether it reduces the search or not.

However, some experiments were done which suggest that it reduces the search.

In these experiments the choice of i_{t+1} according to 2.5 (heuristic) and the choice $i_{t+1} := t+1$ (natural) were compared.

In one case the coefficients of branching were

level	3	4	5	6	7	8	...
heuristical	3	5	20	25	1	1	...
natural	3	20	25	5	1	1	...

In another case the sizes of the trees hanging at some (not too far) advanced vertex of the exhaustion tree were compared. This vertex was fixed and then the remaining numbers were exhausted in the natural and heuristic order. In all these cases the heuristic approach generated trees which were several times smaller than those generated by the natural approach.

7. Possible modifications.

7.1. It is possible to use the results of "Fixation" (cf., 2.4) also before "Break-down" in the same manner as they were used in 3.4. We did not experiment with this possibility.

7.2. It is also possible to use the canonization algorithm not only at the end points of our tree, but at every vertex. However, it is not clear whether it will make the algorithm work faster. Indeed, the canonization algorithm is quite bulky and we already have at least partial canonization (cf., 2.3, 2.5, 2.6, 2.8). There were experiments with this approach in the case $n = 29$, $n_1 = 14$, $a_{11}^1 = 7$, $a_{11}^2 = 6$. But the tree in this case still was too large to be handled by computer (cf. [Ar2]). On the other hand sometimes (when one has a lot of computer time) only the storage space matters.

8. About realizations. Several programs, based on different modifications of the above algorithm were written. One of them was written for a computer M-20 and all others for a computer ICL, System 4-70. The results coincided.

U. TABLES OF STRONGLY REGULAR GRAPHS WITH n VERTICES, $10 \leq n \leq 28$

All strongly regular graphs constructed by the algorithm described in Section T are given in the tables below. Also some information on these graphs is given.

The information about the graphs is arranged as follows. The upper line is

$$n = \quad , n_1 = \quad , \#$$

which shows the number of vertices of the graph ("n"), its degree ("n_1") and its number among the graphs with the same n and n_1.

Below this line the connection table of the corresponding graph is given. The column

"VER"

gives the numeration of the vertices of the graph.

The column

"TYPE"

indicates the number of the canonical form of the neighbour graph of the corresponding vertex.

Under the title

"NEIGHBOURS"

the canonical numeration of the vertices of the neighbour graph of the corresponding vertex is given.

Below the connection table and after the word

"ORBITS"

the nontrivial orbits of the automorphism group of the corresponding graph are given. In the case when this group acts transitively, it is written:

"ORBITS : TRANSITIVE"

In the case when this group is trivial, it is written:

ORBITS : n POINTS"

(where n is the number of vertices of the given graph).

The next line is

"NUMBER OF DIFFERENT NEIGHBOUR TYPES = b"

and this means that our graph has b non-isomorphic neighbour graphs.

In the cases when the pictures of the neighbour graphs are not too complicated, they are given under the heading

"NEIGHBOUR GRAPH"

The 15 graphs with $n = 25$, $n_1 = 12$ are called the 25-family, and the 10 graphs with $n = 26$, $n_1 = 10$ are called the 26-family. For these families there are tables

"MULTIPLICITY OF THE NEIGHBOUR TYPES IN n-FAMILY"

The (i,j)-entry of these tables is the multiplicity of the i-th type of the neighbour graphs in the j-th strongly regular graph.

More heuristical information on the 25- and 26-families is discussed in the next Section.

All graphs with $n = 28$, $n_1 = 12$ are known (cf. [Ch. 2]). All graphs of the 25- and 26-families were independently constructed by A. J. L. Paulus [Pa 1] under the guidance of J. J. Seidel (cf. [Se 5]). However, his algorithm does not guarantee that the constructed families exhaust all strongly regular graphs with given parameters. Our algorithm as it was already indicated constructs complete families. Our results were announced in [Ro 1], [Ro 2], [Ar 2].

All other graphs in our tables have transitive automorphism groups.

$n = 10,\ n_1 = 3,\ \#1$

VER	TYPE	NEIGHBOURS		
1	1	2	3	4
2	1	1	5	6
3	1	1	7	8
4	1	1	9	10
5	1	2	7	9
6	1	2	8	10
7	1	3	5	10
8	1	3	6	9
9	1	4	5	8
10	1	4	6	7

ORBITS: TRANSITIVE

NUMBER OF DIFFERENT NEIGHBOUR TYPES = 1

NEIGHBOUR GRAPH

1 2 3
. . .

$n = 13,\ n_1 = 6,\ \#1$

VER	TYPE	NEIGHBOURS					
1	1	2	3	4	5	6	7
2	1	1	3	4	8	9	10
3	1	1	2	5	8	12	11
4	1	1	2	6	9	11	13
5	1	1	3	7	12	13	9
6	1	1	4	7	11	10	12
7	1	1	5	6	13	10	8
8	1	2	3	10	11	7	13
9	1	2	4	10	13	12	5
10	1	2	8	9	7	12	6
11	1	3	8	12	13	6	4
12	1	3	5	11	9	6	10
13	1	4	9	11	5	8	7

ORBITS: TRANSITIVE

NUMBER OF DIFFERENT NEIGHBOUR TYPES = 1

NEIGHBOUR GRAPH

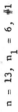

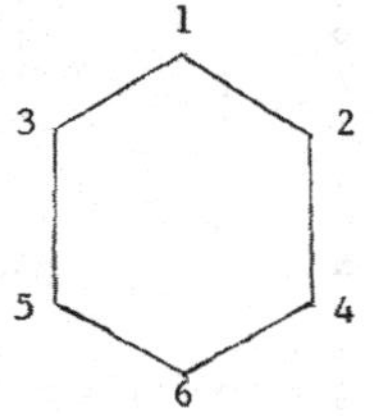

$n = 15,\ n_1 = 6,\ \#1$

VER	TYPE	NEIGHBOURS					
1	1	2	3	4	5	6	7
2	1	1	3	8	10	9	11
3	1	1	2	12	15	13	14
4	1	1	5	8	12	9	13
5	1	1	4	10	15	11	14
6	1	1	7	8	14	9	15
7	1	1	6	10	13	11	12
8	1	2	10	4	12	6	14
9	1	2	11	4	13	6	15
10	1	2	8	5	15	7	13
11	1	2	9	5	14	7	12
12	1	3	15	4	8	7	11
13	1	3	14	4	9	7	10
14	1	3	13	5	11	6	8
15	1	3	12	5	10	6	9

ORBITS: TRANSITIVE

NUMBER OF DIFFERENT NEIGHBOUR TYPES = 1

NEIGHBOUR GRAPH

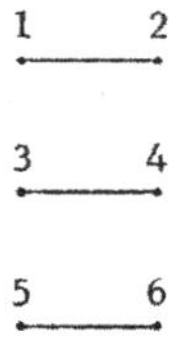

$n = 16,\ n_1 = 5,\ \#1$

VER	TYPE	NEIGHBOURS				
1	1	2	3	4	5	6
2	1	1	7	8	9	10
3	1	1	7	11	12	13
4	1	1	8	11	14	15
5	1	1	9	12	14	16
6	1	1	10	13	15	16
7	1	2	3	14	15	16
8	1	2	4	12	13	16
9	1	2	5	11	13	15
10	1	2	6	11	12	14
11	1	3	4	9	10	16
12	1	3	5	8	10	15
13	1	3	6	8	9	14
14	1	4	5	7	10	13
15	1	4	6	7	9	12
16	1	5	6	7	8	11

ORBITS: TRANSITIVE

NUMBER OF DIFFERENT NEIGHBOUR TYPES = 1

NEIGHBOUR GRAPH

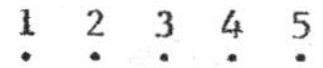

$n = 16,\ n_1 = 6,$ #1

VER	TYPE	NEIGHBOURS					
1	1	2	3	4	5	6	7
2	1	1	3	4	8	9	10
3	1	1	2	4	11	12	13
4	1	1	2	3	14	15	16
5	1	1	6	7	8	11	14
6	1	1	5	7	9	12	15
7	1	1	5	6	10	13	16
8	1	2	9	10	5	11	14
9	1	2	8	10	6	12	15
10	1	2	8	9	7	13	16
11	1	3	12	13	5	8	14
12	1	3	11	13	6	9	15
13	1	3	11	12	7	10	16
14	1	4	15	16	5	8	11
15	1	4	14	16	6	9	12
16	1	4	14	15	7	10	13

ORBITS : TRANSITIVE

NUMBER OF DIFFERENT NEIGHBOUR TYPES = 1

$n = 16,\ n_1 = 6,$ #2

VER	TYPE	NEIGHBOURS					
1	2	2	3	4	5	6	7
2	2	1	3	4	8	9	10
3	2	1	2	5	8	11	12
4	2	1	2	6	9	14	13
5	2	1	3	7	11	15	13
6	2	1	4	7	14	16	12
7	2	1	5	6	15	16	10
8	2	2	3	10	12	15	14
9	2	2	4	10	13	16	11
10	2	2	8	9	15	16	7
11	2	3	5	12	13	16	9
12	2	3	8	11	14	16	6
13	2	4	9	14	11	15	5
14	2	4	6	13	12	15	8
15	2	5	7	13	10	14	8
16	2	6	7	12	10	11	9

ORBITS : TRANSITIVE

NUMBER OF DIFFERENT NEIGHBOUR TYPES = 1

NEIGHBOUR GRAPHS

#1

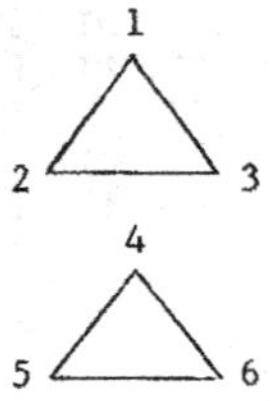

#2

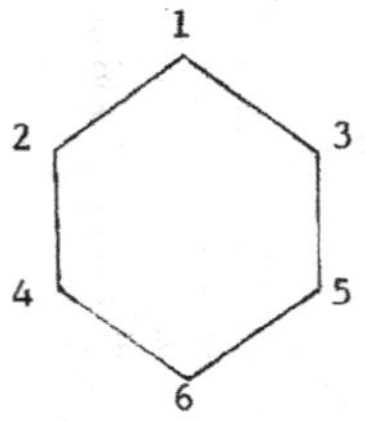

$n = 17,\ n_1 = 8,\ \#1$

VER	TYPE	NEIGHBOURS							
1	1	2	3	4	5	6	7	8	9
2	1	1	4	3	5	10	12	11	13
3	1	1	6	2	7	11	14	10	15
4	1	1	8	2	6	10	17	12	16
5	1	1	2	7	9	12	13	14	16
6	1	1	9	3	4	11	16	14	17
7	1	1	3	5	8	14	15	12	17
8	1	1	7	4	9	17	15	10	13
9	1	1	5	6	8	16	13	11	15
10	1	2	13	3	4	14	8	15	16
11	1	2	12	3	13	15	17	6	9
12	1	2	4	5	11	16	17	7	15
13	1	2	5	10	11	14	9	8	17
14	1	3	6	7	10	17	16	5	13
15	1	3	7	10	11	8	12	16	9
16	1	4	10	6	12	14	15	9	5
17	1	4	12	6	8	11	7	14	13

ORBITS: TRANSITIVE

NUMBER OF DIFFERENT NEIGHBOUR TYPES = 1

NEIGHBOUR GRAPH

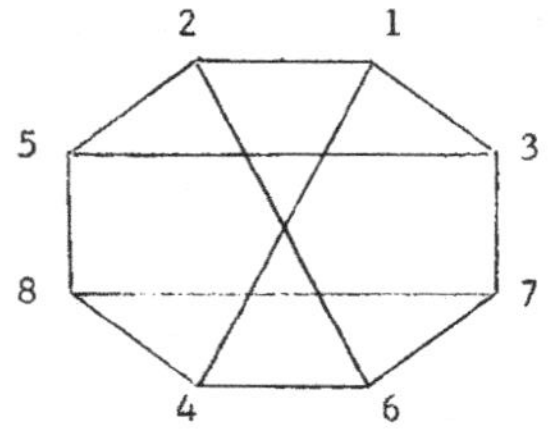

$n = 21,\ n_1 = 10,\ \#1$

VER	TYPE	NEIGHBOURS									
1	1	2	3	4	5	6	7	8	9	10	11
2	1	1	3	4	5	6	7	12	13	14	15
3	1	1	2	4	5	6	8	12	16	17	18
4	1	1	2	3	5	6	9	13	16	19	20
5	1	1	2	3	4	6	10	14	17	19	21
6	1	1	2	3	4	5	11	15	18	20	21
7	1	1	8	9	10	11	2	12	13	14	15
8	1	1	7	9	10	11	3	12	16	17	18
9	1	1	7	8	10	11	4	13	16	19	20
10	1	1	7	8	9	11	5	14	17	19	21
11	1	1	7	8	9	10	6	15	18	20	21
12	1	2	7	13	14	15	3	8	16	17	18
13	1	2	7	12	14	15	4	9	16	19	20
14	1	2	7	12	13	15	5	10	17	19	21
15	1	2	7	12	13	14	6	11	18	20	21
16	1	3	8	12	17	18	4	9	13	19	20
17	1	3	8	12	16	18	5	10	14	19	21
18	1	3	8	12	16	17	6	11	15	20	21
19	1	4	9	13	16	20	5	10	14	17	21
20	1	4	9	13	16	19	6	11	15	18	21
21	1	5	10	14	17	19	6	11	15	18	20

ORBITS: TRANSITIVE

NUMBER OF DIFFERENT NEIGHBOUR TYPES = 1

$n = 25,\ n_1 = 8,$ #1

VER	TYPE	NEIGHBOURS							
1	1	2	3	4	5	6	7	8	9
2	1	1	3	4	5	10	11	12	13
3	1	1	2	4	5	14	15	16	17
4	1	1	2	3	5	18	19	20	21
5	1	1	2	3	4	22	23	24	25
6	1	1	7	8	9	10	14	18	22
7	1	1	6	8	9	11	15	19	23
8	1	1	6	7	9	12	16	20	24
9	1	1	6	7	8	13	17	21	25
10	1	2	11	12	13	6	14	18	22
11	1	2	10	12	13	7	15	19	23
12	1	2	10	11	13	8	16	20	24
13	1	2	10	11	12	9	17	21	25
14	1	3	15	16	17	6	10	18	22
15	1	3	14	16	17	7	11	19	23
16	1	3	14	15	17	8	12	20	24
17	1	3	14	15	16	9	13	21	25
18	1	4	19	20	21	6	10	14	22
19	1	4	18	20	21	7	11	15	23
20	1	4	18	19	21	8	12	16	24
21	1	4	18	19	20	9	13	17	25
22	1	5	23	24	25	6	10	14	18
23	1	5	22	24	25	7	11	15	19
24	1	5	22	23	25	8	12	16	20
25	1	5	22	23	24	9	13	17	21

ORBITS: TRANSITIVE

NUMBER OF DIFFERENT NEIGHBOUR TYPES = 1

NEIGHBOUR GRAPH

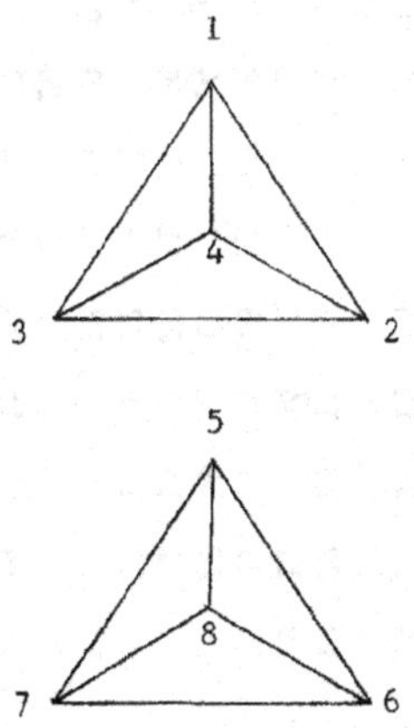

$n = 25,\ n_1 = 12$, #1

VER	TYPE	NEIGHBOURS											
1	1	2	3	4	5	6	7	8	9	10	11	12	13
2	1	1	3	7	4	5	6	16	18	19	14	15	17
3	2	1	2	6	4	5	7	21	23	25	20	22	24
4	3	1	8	10	9	2	3	14	16	15	21	22	20
5	4	20	24	17	11	3	18	8	2	14	1	12	23
6	3	23	21	3	25	19	9	15	11	2	1	17	13
7	3	2	18	16	19	1	3	12	10	13	25	22	24
8	3	10	1	9	4	24	25	11	5	19	23	16	14
9	1	19	23	8	6	22	18	1	4	10	13	17	20
10	5	24	25	17	7	8	15	21	18	9	1	12	4
11	6	8	24	19	5	16	1	15	20	6	12	13	21
12	7	14	22	23	13	15	5	7	21	18	11	1	10
13	8	9	17	6	20	1	22	25	14	11	16	7	12
14	7	12	22	23	15	13	5	4	25	8	17	2	16
15	8	19	11	6	24	2	22	21	12	17	10	4	14
16	5	20	21	11	4	18	13	25	8	19	2	14	7
17	6	18	20	9	5	10	2	13	24	6	14	15	25
18	3	16	2	19	7	20	21	17	5	9	23	10	12
19	1	9	23	18	6	22	8	2	7	16	15	11	24
20	9	21	16	4	11	18	3	13	9	22	5	24	17
21	10	20	16	4	18	11	3	25	10	15	12	23	6
22	11	12	14	13	15	23	7	4	9	20	19	24	3
23	2	9	19	6	8	22	18	25	3	21	14	5	12
24	9	25	10	7	17	8	3	15	19	22	5	20	11
25	10	24	10	7	8	17	3	21	16	13	14	23	6

$n = 25,\ n_1 = 12$, #2

VER	TYPE	NEIGHBOURS											
1	2	2	3	4	5	6	7	8	9	10	11	12	13
2	1	1	3	5	4	7	6	14	18	17	16	15	19
3	1	1	2	7	4	5	6	22	24	25	20	21	23
4	12	16	14	2	21	8	15	22	3	1	9	20	10
5	5	23	24	3	8	17	20	12	1	2	14	11	18
6	13	15	2	19	17	11	9	1	3	23	25	13	21
7	5	19	18	2	10	25	16	12	1	3	22	13	24
8	3	4	1	9	10	14	16	11	5	25	24	19	23
9	14	8	10	4	1	24	25	17	18	15	21	12	6
10	3	4	1	9	8	22	20	13	7	17	18	23	19
11	15	15	20	19	12	22	6	5	25	8	13	1	14
12	16	16	7	13	24	21	15	1	18	11	9	5	20
13	15	21	16	23	12	14	6	7	17	10	11	1	22
14	5	16	4	21	8	2	13	22	25	18	5	11	17
15	17	11	20	19	22	12	6	4	16	2	24	17	9
16	18	14	4	21	2	8	13	15	23	12	19	7	24
17	3	18	2	5	14	9	10	15	6	24	23	22	13
18	5	19	7	10	25	2	20	12	9	17	14	21	5
19	19	7	18	2	10	25	16	20	6	15	8	23	11
20	18	15	11	22	12	19	4	5	10	3	18	21	23
21	17	13	16	23	14	12	6	4	20	3	18	25	9
22	5	20	4	15	10	3	11	14	17	24	7	13	25
23	19	5	24	3	8	17	20	16	6	21	10	19	13
24	5	23	5	8	17	3	16	12	9	25	22	15	7
25	3	24	3	7	22	9	8	21	6	18	19	14	11

#1 ORBITS: (1,2)(4,7)(8,18)(9,19)(10,16)(11,17)(12,14)(13,15)(20,24)(21,25)
NUMBER OF DIFFERENT NEIGHBOUR TYPES = 11

#2 ORBITS: (2,3)(5,7)(17,25)(14,22)(18,24)(15,21)(19,23)(16,20)(11,13)(8,10)
NUMBER OF DIFFERENT NEIGHBOUR TYPES = 12

$n = 25,\ n_1 = 12,$ #3

VER	TYPE	NEIGHBOURS											
1	2	2	3	4	5	6	7	8	9	10	11	12	13
2	1	1	3	5	4	7	6	14	18	17	16	15	19
3	1	1	2	7	4	6	5	22	25	24	21	20	23
4	3	1	8	9	10	2	3	16	14	15	22	20	21
5	5	11	8	1	24	14	12	23	3	2	17	20	18
6	20	25	9	17	23	13	3	15	1	2	21	19	11
7	3	22	24	3	25	18	12	19	10	2	1	16	13
8	21	11	5	24	1	14	19	23	10	25	4	9	16
9	22	4	1	10	8	15	22	12	6	17	18	25	23
10	3	1	4	9	8	13	7	20	21	17	18	24	19
11	15	5	8	1	24	14	12	19	13	6	22	15	21
12	19	20	16	23	13	15	5	7	9	18	11	1	22
13	6	7	12	1	16	10	25	11	20	6	14	17	21
14	5	21	22	11	4	18	13	25	8	5	2	16	17
15	8	6	2	19	17	11	9	16	4	24	20	12	22
16	15	12	20	13	15	23	7	4	25	14	19	2	8
17	5	20	24	10	15	5	13	25	9	18	2	6	14
18	3	17	2	14	5	10	9	19	7	21	22	23	12
19	23	6	11	15	21	2	23	24	8	16	18	10	7
20	9	12	16	23	15	13	5	4	21	3	17	24	10
21	7	14	22	11	4	18	13	3	19	6	10	20	23
22	10	21	14	4	18	11	3	25	9	15	12	7	24
23	24	12	5	20	18	16	9	3	8	21	19	25	6
24	10	20	17	5	10	15	3	25	8	11	19	7	22
25	25	3	22	24	7	6	23	9	14	17	8	13	16

$n = 25,\ n_1 = 12,$ #4

VER	TYPE	NEIGHBOURS											
1	1	2	3	4	5	6	7	8	9	10	11	12	13
2	1	1	3	5	4	7	6	14	18	17	16	15	19
3	1	1	2	7	4	5	6	22	24	25	20	21	23
4	5	16	14	2	21	8	15	22	3	1	10	20	9
5	3	2	14	17	18	1	3	11	8	12	24	20	23
6	26	15	2	19	17	11	9	1	3	25	23	21	13
7	3	22	24	3	25	13	18	12	16	1	2	10	19
8	5	23	24	16	5	9	19	25	14	4	1	11	10
9	3	8	1	10	4	23	24	13	6	18	17	22	15
10	3	1	4	9	8	12	7	21	20	17	18	25	19
11	15	15	20	19	12	22	6	5	25	8	13	1	14
12	6	7	13	1	16	10	24	11	21	5	15	17	20
13	9	18	22	9	14	7	23	11	1	6	16	21	12
14	5	16	4	21	8	2	13	22	25	17	5	11	18
15	7	20	11	22	19	12	4	6	24	9	16	2	17
16	18	14	4	21	2	8	13	15	23	12	19	7	24
17	5	21	25	10	6	14	12	24	9	18	2	15	5
18	5	13	22	9	7	14	23	20	10	17	2	19	5
19	20	11	6	15	25	20	8	2	23	16	10	7	18
20	27	15	11	22	12	19	4	5	18	3	10	21	23
21	9	13	16	23	14	12	6	4	20	3	17	25	10
22	21	13	18	14	7	9	11	20	25	4	24	3	15
23	17	24	8	5	16	9	3	19	18	20	13	21	6
24	10	23	8	5	9	16	3	25	17	12	15	22	7
25	28	21	17	6	10	14	3	24	19	11	8	7	22

#3 { ORBITS: 25 POINTS
NUMBER OF DIFFERENT NEIGHBOUR TYPES = 17

#4 { ORBITS: 25 POINTS
NUMBER OF DIFFERENT NEIGHBOUR TYPES = 15

$n = 25,\ n_1 = 15$, #5

VER	TYPE	NEIGHBOURS											
1	1	2	3	4	5	6	7	8	9	10	11	12	13
2	1	1	3	5	4	7	6	14	18	17	16	15	19
3	1	1	2	7	4	6	5	22	25	24	21	20	23
4	5	21	22	3	14	10	20	15	2	1	8	16	9
5	5	11	8	1	24	14	12	23	3	2	18	20	17
6	23	9	25	17	13	15	1	23	3	2	11	21	19
7	5	19	18	2	10	24	16	13	1	3	22	12	25
8	5	11	5	24	14	1	19	23	25	9	4	16	10
9	3	8	1	10	4	25	24	6	13	17	18	15	20
10	3	4	1	9	8	22	21	12	7	17	18	23	19
11	15	5	8	1	24	14	12	19	13	6	22	15	21
12	15	20	16	23	13	15	5	7	17	10	11	1	22
13	29	16	7	12	25	20	14	1	18	11	9	6	21
14	3	5	2	17	18	8	11	16	4	25	22	13	21
15	29	20	4	16	9	12	24	2	22	19	17	6	11
16	15	12	20	13	15	23	7	4	25	14	19	2	8
17	3	18	2	14	5	9	10	6	15	25	22	23	12
18	5	19	7	10	24	2	21	13	9	17	5	20	14
19	15	7	18	2	10	24	16	21	15	6	8	23	11
20	15	16	12	15	23	13	4	5	9	24	21	3	18
21	15	4	22	3	14	10	20	11	23	6	18	13	19
22	5	21	4	14	10	3	11	15	17	25	7	12	24
23	29	12	5	20	17	16	10	3	8	21	25	6	19
24	3	7	3	25	22	18	19	20	5	9	8	15	11
25	3	22	3	24	7	17	14	6	23	9	8	13	16

$n = 25,\ n_1 = 12$, #6

VER	TYPE	NEIGHBOURS											
1	1	2	3	4	5	6	7	8	9	10	11	12	13
2	2	1	3	6	4	7	5	15	19	17	16	14	18
3	1	1	2	5	4	6	7	20	23	24	21	22	25
4	14	1	10	9	8	2	3	15	20	16	21	14	22
5	14	3	24	23	20	1	2	8	17	11	18	12	14
6	22	15	19	2	17	11	23	25	13	1	3	9	21
7	13	10	18	12	25	1	19	2	16	22	13	3	24
8	2	17	24	9	5	22	19	1	4	10	11	14	25
9	22	1	4	8	10	13	6	21	16	24	17	23	18
10	1	15	20	4	12	19	23	1	8	9	7	18	25
11	26	15	16	23	22	6	12	13	14	25	5	8	1
12	13	1	13	7	11	10	5	20	21	22	18	15	17
13	26	1	11	6	12	9	7	16	14	21	19	20	24
14	30	2	16	18	4	5	19	11	13	21	25	8	20
15	1	10	20	19	4	23	12	2	6	17	16	22	11
16	26	24	13	9	7	23	22	11	14	18	4	2	15
17	1	8	24	19	5	9	22	2	6	15	18	12	21
18	26	10	25	23	7	9	12	21	14	16	5	2	17
19	14	15	17	6	2	10	20	8	24	25	13	7	14
20	2	10	15	23	4	12	19	3	5	24	21	14	13
21	26	17	18	9	12	6	22	25	14	13	4	20	3
22	13	3	25	7	21	24	4	8	11	12	16	17	15
23	22	3	5	20	24	25	6	11	18	10	15	9	16
24	1	8	17	5	9	22	19	23	3	20	16	13	7
25	26	3	21	6	22	23	7	18	14	11	19	8	10

#5 ORBITS: (1,2,3)(4,5,7)(10,14,24)(8,22,18)(9,17,25)(12,20,16)(13,15,23)(11,19,21)
NUMBER OF DIFFERENT NEIGHBOUR TYPES = 6

#6 ORBITS: (1,15,17,24,10,3)(2,20,8)(5,19,4)(6,23,9)(7,22,12)(11,16,18,21,25,13)
NUMBER OF DIFFERENT NEIGHBOUR TYPES = 7

$n = 25,\ n_1 = 12$, #7

VER	TYPE	NEIGHBOURS											
1	1	2	3	4	5	6	7	8	9	10	11	12	13
2	2	1	3	6	4	7	5	15	19	17	16	14	18
3	1	1	2	5	4	7	6	20	24	23	22	21	25
4	3	10	8	1	9	21	15	22	14	3	2	20	16
5	5	11	8	1	14	24	12	17	2	3	23	18	20
6	3	17	15	2	19	25	9	21	11	3	1	23	13
7	31	12	1	13	10	22	18	3	2	19	16	24	25
8	5	11	5	24	14	1	25	17	19	10	4	22	9
9	5	25	17	18	6	8	16	20	23	10	1	13	4
10	1	19	24	8	7	15	23	1	4	9	12	18	21
11	4	5	8	1	24	14	12	25	13	6	16	15	21
12	27	20	17	22	5	15	13	18	21	7	11	1	10
13	20	22	7	12	19	20	14	1	16	11	23	6	9
14	13	18	5	2	23	16	21	11	8	19	4	13	22
15	3	17	2	19	6	20	12	16	4	24	10	11	21
16	20	18	7	25	2	9	14	24	13	11	4	15	20
17	32	12	20	5	15	22	18	9	2	8	19	6	25
18	24	10	21	23	12	9	7	14	25	5	17	16	2
19	2	10	24	15	8	7	23	17	2	6	22	14	13
20	5	12	17	22	15	5	13	9	4	3	24	16	23
21	16	3	25	6	22	23	4	11	18	15	14	12	10
22	24	3	21	4	25	20	7	14	12	8	17	13	19
23	3	3	5	24	20	21	6	18	14	10	19	9	13
24	1	10	19	23	7	8	15	3	5	20	25	16	11
25	27	9	17	18	8	6	16	22	21	7	11	24	3

$n = 25,\ n_1 = 12$, #8

VER	TYPE	NEIGHBOURS											
1	1	2	3	4	5	6	7	8	9	10	11	12	13
2	1	1	3	6	4	5	7	15	17	19	14	16	18
3	1	1	2	5	4	6	7	20	23	24	21	22	25
4	3	10	8	1	9	15	20	14	22	2	3	16	21
5	3	3	20	24	23	1	2	11	12	8	17	14	18
6	3	17	15	2	19	9	21	13	23	1	3	11	25
7	8	22	24	13	3	12	25	16	19	1	2	18	10
8	1	17	24	9	5	19	22	1	10	4	11	14	25
9	3	10	1	8	4	23	18	13	6	24	17	16	21
10	1	15	20	4	12	19	23	1	8	9	7	18	25
11	8	1	12	5	13	8	6	20	21	14	16	19	25
12	8	7	1	13	10	22	18	11	5	15	20	21	17
13	23	6	9	23	1	15	11	24	16	14	12	7	22
14	8	15	22	4	13	2	23	8	25	16	11	5	18
15	1	10	20	19	4	23	12	2	6	17	14	22	13
16	8	24	7	19	13	20	9	2	18	11	14	4	21
17	1	8	24	19	5	9	22	2	6	15	18	12	21
18	29	10	25	23	7	9	12	14	21	5	16	2	17
19	3	17	2	15	6	24	8	16	7	20	10	11	25
20	1	10	15	23	4	12	19	3	5	24	21	16	11
21	29	17	18	9	12	6	22	16	25	4	11	20	3
22	8	12	15	13	17	7	21	14	4	24	8	25	3
23	3	3	5	20	24	25	6	18	14	10	15	9	13
24	1	8	17	5	9	22	19	23	3	20	13	16	7
25	29	3	21	6	22	23	7	11	18	19	14	8	10

#7 ORBITS: (1,24)(2,19)(3,10)(4,23)(5,8)(6,15)(9,20)(12,25)(13,16)(18,22)
NUMBER OF DIFFERENT NEIGHBOUR TYPES = 12

#8 ORBITS: (1,24,15)(2,20,8)(3,17,10)(4,19,5)(6,23,9)(7,22,12)(11,16,14)(18,25,21)
NUMBER OF DIFFERENT NEIGHBOUR TYPES = 5

n = 25, n_1 = 12, #9

VER	TYPE	NEIGHBOURS											
1	1	2	3	4	5	6	7	8	9	10	11	12	13
2	1	1	3	6	4	5	7	15	17	19	14	16	18
3	1	1	2	5	4	7	6	20	24	23	22	21	25
4	22	9	8	1	10	16	20	22	14	2	3	15	21
5	22	20	23	3	24	12	17	18	11	1	2	8	14
6	22	17	15	2	19	25	9	13	21	3	1	23	11
7	30	12	1	10	13	18	22	2	3	24	19	16	25
8	22	1	4	10	9	11	5	14	22	19	24	17	25
9	32	16	20	23	4	13	18	17	10	6	1	8	25
10	1	19	24	8	7	15	23	1	4	9	12	18	21
11	31	1	12	5	13	8	6	22	21	14	23	16	19
12	11	17	20	5	15	22	18	13	11	1	21	10	7
13	13	1	11	6	12	9	7	16	14	15	25	20	24
14	31	2	16	4	18	15	5	13	11	21	8	25	24
15	22	2	6	19	17	14	4	21	13	24	10	20	12
16	11	9	20	23	4	13	18	22	11	19	14	2	7
17	32	12	20	5	15	22	18	9	2	8	19	6	25
18	13	2	14	5	16	17	7	25	21	23	12	9	10
19	1	10	24	15	7	23	8	2	6	17	16	22	11
20	32	12	17	15	5	22	13	9	24	4	3	23	16
21	31	3	25	6	22	23	4	18	14	11	15	12	10
22	13	3	21	4	25	20	7	12	11	8	16	17	19
23	22	3	5	24	20	21	6	11	18	10	19	9	16
24	1	10	19	23	7	8	15	3	5	20	25	13	14
25	11	9	17	6	8	18	13	22	21	3	14	24	7

n = 25, n_1 = 12, # 10

VER	TYPE	NEIGHBOURS											
1	33	2	3	4	5	6	7	8	9	10	11	12	13
2	33	1	3	4	5	6	7	14	15	16	17	18	19
3	33	1	2	4	5	6	7	20	21	22	23	24	25
4	34	1	8	9	10	3	2	20	14	21	15	22	16
5	35	18	17	20	2	12	23	8	11	3	1	14	24
6	35	17	19	21	2	11	25	9	13	3	1	15	23
7	35	19	18	22	2	13	24	10	12	3	1	16	25
8	36	9	4	1	10	19	24	14	20	13	5	17	25
9	36	8	4	1	10	19	24	21	15	6	12	18	23
10	36	8	4	1	9	17	25	16	22	11	7	18	23
11	37	16	17	21	10	12	14	6	5	22	1	23	13
12	37	15	18	20	9	13	16	5	7	21	1	24	11
13	37	14	19	22	8	11	15	7	6	20	1	25	12
14	38	2	4	16	15	19	5	8	22	24	11	13	23
15	38	2	4	14	16	18	6	9	20	23	13	12	25
16	38	2	4	15	14	17	7	10	21	25	12	11	24
17	39	5	20	18	8	2	11	25	21	19	10	6	16
18	39	7	22	19	10	2	12	23	20	17	9	5	15
19	39	6	21	17	9	2	13	24	22	18	8	7	14
20	38	3	4	21	22	25	5	8	15	17	12	13	18
21	38	3	4	22	20	24	6	9	16	19	11	12	17
22	38	3	4	20	21	23	7	10	14	18	13	11	19
23	39	6	15	25	9	3	11	18	14	24	10	5	22
24	39	5	14	23	8	3	12	19	16	25	9	7	21
25	39	7	16	24	10	3	13	17	15	23	8	6	20

#9 ORBITS: (1,24,19,10,3,2)(4,23,15,8,6,5)(9,20,17)(11,21,14)(12,25,16)(13,22,18)
NUMBER OF DIFFERENT NEIGHBOUR TYPES = 7

#10 ORBITS: (1)(2,3)(5,7,6)(8,10,9)(11,13,12)(14,22,21,20,16,15)(17,25,24,23,19,18)
NUMBER OF DIFFERENT NEIGHBOUR TYPES = 7

$n = 25,\ n_1 = 12$, #11

VER	TYPE	NEIGHBOURS											
1	33	2	3	4	5	6	7	8	9	10	11	12	13
2	33	1	3	4	5	6	7	14	15	16	17	18	19
3	33	1	2	4	5	6	7	20	21	22	23	24	25
4	40	1	8	9	10	2	3	15	22	16	20	14	21
5	41	11	12	20	14	1	18	23	3	17	2	24	8
6	41	11	13	15	21	1	25	17	2	23	3	19	9
7	41	12	13	22	16	1	19	24	3	18	2	25	10
8	33	15	22	4	13	18	23	1	10	9	17	24	5
9	33	16	20	4	12	17	25	1	8	10	19	23	6
10	33	14	21	4	11	19	24	1	9	8	18	25	7
11	41	5	12	20	1	14	18	13	21	25	10	6	15
12	41	5	11	14	1	20	23	13	16	19	9	7	22
13	41	6	11	21	1	15	17	12	22	24	8	7	16
14	33	10	21	4	11	19	24	15	2	16	12	5	23
15	33	8	22	4	13	18	23	16	2	14	11	6	25
16	33	9	20	4	12	17	25	14	2	15	13	7	24
17	41	5	18	8	2	20	24	19	9	13	16	6	21
18	41	5	17	20	2	8	11	19	22	25	15	7	10
19	41	6	17	9	2	21	23	18	10	12	14	7	22
20	33	9	16	4	12	17	25	22	21	3	5	11	18
21	33	10	14	4	11	19	24	20	22	3	6	13	17
22	33	8	15	4	13	18	23	21	20	3	7	12	19
23	41	5	24	14	3	8	12	25	15	19	22	6	9
24	41	5	23	8	3	14	17	25	10	13	21	7	16
25	41	6	23	15	3	9	11	24	16	18	20	7	10

$n = 25,\ n_1 = 12$, #12

VER	TYPE	NEIGHBOURS											
1	41	2	3	4	5	6	7	8	9	10	11	12	13
2	42	1	3	4	5	6	7	14	15	16	17	18	19
3	42	1	2	5	4	6	8	14	20	23	21	22	24
4	37	16	22	15	9	2	20	10	3	25	1	21	7
5	37	23	18	20	11	3	15	12	2	25	1	17	8
6	35	17	16	24	2	11	21	13	9	3	1	19	23
7	42	1	10	4	11	13	2	14	25	20	24	18	19
8	42	1	12	5	9	13	3	14	25	15	19	22	24
9	35	18	16	22	12	11	19	4	6	8	1	21	25
10	42	1	7	13	4	11	12	14	15	23	22	21	17
11	35	22	18	23	9	10	24	5	7	6	1	25	17
12	42	1	8	13	5	9	10	14	20	16	18	17	21
13	37	16	24	20	6	12	15	7	8	23	1	19	10
14	43	2	7	19	18	17	3	24	10	21	8	12	22
15	44	2	16	17	4	5	19	22	13	25	10	8	23
16	39	6	24	17	13	2	9	20	22	15	12	4	18
17	38	10	12	21	14	15	11	5	16	25	6	2	24
18	38	2	7	14	19	5	16	20	11	12	22	23	9
19	38	2	7	18	14	15	6	13	25	23	9	8	21
20	44	3	23	21	5	4	24	18	13	25	12	7	16
21	38	10	12	14	17	4	23	20	9	3	19	25	6
22	38	3	8	14	24	4	23	15	9	10	18	16	11
23	39	6	19	21	13	3	11	15	18	20	10	5	22
24	38	3	8	22	14	20	6	13	25	16	11	7	17
25	39	9	21	19	4	8	11	20	17	15	7	5	24

#11 { ORBITS: (1,22,21,20,16,15,14,10,9,8,3,2)(5,25,24,23,19,18,17,13,12,11,7,6)
NUMBER OF DIFFERENT NEIGHBOUR TYPES = 3

#12 { ORBITS: (2,12,10,8,7,3)(4,13,5)(6,11,9)(15,20)(16,25,23)(17,24,22,21,19,18)
NUMBER OF DIFFERENT NEIGHBOUR TYPES = 8

$n = 25,\ n_1 = 12$, #13

VER	TYPE	NEIGHBOURS											
1	15	2	3	4	5	6	7	8	9	10	11	12	13
2	10	1	3	4	5	6	7	14	15	16	17	18	19
3	12	1	2	5	6	4	8	14	23	20	21	24	22
4	10	16	15	25	2	10	21	22	7	9	1	3	20
5	3	25	11	23	15	8	18	1	12	3	20	2	17
6	20	11	21	19	13	23	1	16	3	2	9	17	24
7	32	9	25	4	18	24	1	11	2	22	14	19	12
8	20	25	21	18	16	5	24	13	3	14	12	10	1
9	7	25	7	24	4	18	23	1	17	6	20	10	13
10	23	1	13	8	9	12	4	14	15	16	20	23	19
11	28	12	7	1	19	22	5	25	6	13	21	23	15
12	7	11	7	1	22	19	5	24	10	8	20	17	16
13	15	22	15	11	14	17	21	10	6	1	18	8	9
14	10	24	3	22	23	8	7	2	15	13	10	19	18
15	32	13	22	14	11	17	10	4	23	2	5	25	16
16	15	4	15	25	2	10	21	17	8	24	19	6	12
17	23	5	18	20	2	12	15	9	13	22	16	6	24
18	3	19	2	7	14	20	21	17	5	9	25	13	8
19	3	18	2	7	14	21	20	6	16	11	12	23	10
20	3	3	4	21	22	23	5	10	9	19	18	12	17
21	22	4	3	20	22	16	25	8	6	19	18	11	13
22	10	13	15	11	14	17	21	4	7	12	24	3	20
23	3	11	15	5	25	19	6	10	14	20	3	9	24
24	15	14	3	22	23	8	7	6	12	17	25	9	16
25	28	9	7	24	4	18	23	11	16	8	21	15	5

$n = 25,\ n_1 = 12$, #14

VER	TYPE	NEIGHBOURS											
1	40	2	3	4	5	6	7	8	9	10	11	12	13
2	34	1	3	4	5	6	7	14	15	16	17	18	19
3	34	1	2	4	5	9	8	14	15	20	21	23	22
4	34	1	2	3	5	10	11	17	16	20	21	25	24
5	34	1	2	3	4	13	12	18	19	22	23	25	24
6	34	1	8	10	12	7	2	22	16	20	18	24	14
7	34	1	9	11	13	2	6	19	20	15	24	17	22
8	34	1	6	10	12	3	9	22	16	15	25	21	19
9	34	1	7	11	13	8	3	19	20	16	23	25	14
10	34	1	6	8	12	11	4	18	20	15	25	23	17
11	34	1	7	9	13	4	10	24	15	16	23	21	18
12	34	1	6	8	10	5	13	24	14	19	21	23	17
13	34	1	7	9	11	12	5	17	22	14	25	21	18
14	43	2	6	16	18	17	3	20	12	21	9	13	23
15	43	2	7	19	17	18	3	22	11	23	8	10	21
16	43	2	6	18	14	19	4	24	8	25	11	9	21
17	43	2	7	15	19	14	4	20	13	21	10	12	25
18	43	2	6	14	16	15	5	22	10	23	13	11	25
19	43	2	7	17	15	16	5	24	9	25	12	8	23
20	43	3	9	23	14	22	4	25	7	24	10	6	17
21	43	3	8	15	22	14	4	16	12	17	11	13	24
22	43	3	8	21	15	20	5	25	6	24	13	7	18
23	43	3	9	14	20	15	5	19	11	18	12	10	24
24	43	4	11	21	16	20	5	23	7	22	12	6	19
25	43	4	10	17	20	16	5	18	8	19	13	9	22

#13 ORBITS: (13,16)(22,4)(11,25)(14,2)(17,10)(6,8)(1,24)(18,19)(9,12)(20)(23,5)
NUMBER OF DIFFERENT NEIGHBOUR TYPES = 10

#14 ORBITS: (2,13,12,11,10,9,8,7,6,5,4,3)(14,25,24,23,22,21,20,19,18,17,16,15)
NUMBER OF DIFFERENT NEIGHBOUR TYPES = 3

$n = 25,\ n_1 = 12,\ \#15$

VER	TYPE	NEIGHBOURS											
1	45	2	3	4	5	6	7	8	9	10	11	12	13
2	45	1	4	3	5	6	7	16	17	14	15	18	19
3	45	1	5	2	4	8	9	22	23	15	14	20	21
4	45	1	3	2	5	10	11	21	20	17	16	25	24
5	45	1	2	3	4	12	13	19	18	23	22	24	25
6	45	1	10	8	12	2	9	14	17	16	22	18	24
7	45	1	11	9	13	2	8	15	16	17	23	19	25
8	45	1	12	6	10	3	7	15	23	22	16	20	25
9	45	1	13	7	11	3	6	14	22	23	17	21	24
10	45	1	8	6	12	4	13	25	20	17	14	21	19
11	45	1	9	7	13	4	12	24	21	16	15	20	18
12	45	1	6	8	10	5	11	24	18	23	15	19	21
13	45	1	7	9	11	5	10	25	19	22	14	18	20
14	45	2	16	6	18	3	19	23	20	9	10	21	13
15	45	2	17	7	19	3	18	22	21	8	11	20	12
16	45	2	18	6	14	4	7	11	25	24	8	20	23
17	45	2	19	7	15	4	6	10	24	25	9	21	22
18	45	2	6	14	16	5	15	22	12	13	21	25	11
19	45	2	7	15	17	5	14	23	13	12	20	24	10
20	45	3	8	15	22	4	14	16	10	11	19	24	13
21	45	3	9	14	23	4	15	17	11	10	18	25	12
22	45	3	20	8	15	5	9	13	24	25	6	18	17
23	45	3	21	9	14	5	8	12	25	24	7	19	16
24	45	4	20	11	16	5	17	19	22	12	9	23	6
25	45	4	21	10	17	5	16	18	23	13	8	22	7

ORBITS: TRANSITIVE

NUMBER OF DIFFERENT NEIGHBOUR TYPES = 1

MULTIPLICITY OF THE NEIGHBOUR TYPES IN 25-FAMILY

TYPE \ #	1	2	3	4	5	6	7	8	9	13	10	11	12	14	15
1	4	2	2	3	3	6	4	9	8						
2	2	1	1			3	2								
3	5	4	4	4	6		4	6		5					
4	1						1								
5	2	6	3	5	6		4								
6	2		1	1											
7	2		1	1						2					
8	2		1					6							
9	2		1	2											
10	2		2	1						4					
11	1								3						
12		1								1					
13		1				3	1		3						
14		1				3									
15		2	2	1	6					4					
16		1					1								
17		2		1											
18		2		1											
19		2	1												
20			1	1			2			2					
21			1	1											
22			1			3			6	1					
23			1		1			1		2					
24			1				2								
25			1												
26				1		6									
27				1			2								
28				1						2					
29					3			3							
30						1			1						
31							1		3						
32							1		3	2					
33											3	12			
34											1			12	
35											3		3		
36											3				
37											3		3		
38											6		6		
39											6		3		
40												1		1	
41												12	1		
42													6		
43													1	12	
44													2		
45															25

$n = 26,\ n_1 = 10,\ \#1$

VER	TYPE	NEIGHBOURS									
1	1	2	3	4	5	6	7	8	9	10	11
2	2	3	1	4	12	5	13	14	17	15	16
3	3	1	2	4	6	12	18	22	21	19	20
4	4	3	1	2	18	7	13	25	24	23	26
5	4	2	14	15	1	24	20	8	9	23	19
6	4	26	14	25	9	21	8	16	1	3	22
7	4	10	1	11	15	4	21	25	16	23	19
8	4	19	17	25	5	13	6	24	1	10	22
9	4	11	17	18	1	16	20	6	5	23	26
10	3	1	7	11	8	15	12	22	13	20	26
11	2	10	1	7	12	9	21	17	14	18	24
12	5	2	17	3	14	19	11	22	26	23	10
13	3	2	16	17	4	21	8	24	26	20	10
14	2	2	5	15	12	24	25	11	26	21	6
15	4	14	2	5	25	16	20	7	18	22	10
16	4	13	2	17	21	15	9	7	6	22	23
17	2	2	13	16	12	8	9	19	11	25	18
18	4	17	9	11	25	20	24	15	4	3	22
19	3	3	20	21	12	5	7	23	17	8	25
20	1	3	19	21	18	5	13	9	15	26	10
21	3	3	19	20	6	7	13	16	14	11	24
22	6	3	6	12	18	8	16	10	23	24	15
23	6	4	7	24	26	16	19	22	5	9	12
24	4	21	11	14	13	18	5	4	8	22	23
25	1	14	26	6	15	4	8	7	18	19	17
26	3	6	14	25	9	12	4	23	20	10	13

$n = 26,\ n_1 = 10,\ \#2$

VER	TYPE	NEIGHBOURS									
1	1	2	3	4	5	6	7	8	9	10	11
2	2	3	1	4	12	5	13	14	17	15	16
3	3	1	2	4	6	12	18	22	21	20	19
4	1	2	3	1	13	18	7	24	26	23	25
5	3	2	14	15	1	24	19	8	9	23	20
6	4	26	14	25	9	21	8	16	1	3	22
7	1	10	11	1	15	21	4	16	19	23	25
8	4	19	17	25	5	13	6	24	1	10	22
9	3	5	20	23	1	26	16	6	11	18	17
10	3	1	7	11	8	15	12	22	13	26	20
11	2	10	1	7	12	9	21	17	14	18	24
12	7	2	14	17	3	11	25	20	22	10	23
13	3	2	16	17	4	21	8	24	26	20	10
14	2	2	5	15	12	24	26	11	25	21	6
15	4	5	2	14	19	16	26	7	18	22	10
16	4	13	2	17	21	15	9	7	6	22	23
17	2	2	13	16	12	8	9	25	11	19	18
18	4	9	11	17	26	24	19	4	15	22	3
19	3	3	20	21	18	5	7	15	17	8	25
20	2	3	19	21	12	5	13	23	10	9	26
21	3	3	19	20	6	7	13	16	14	11	24
22	6	3	6	12	18	8	16	10	23	24	15
23	3	4	7	25	24	16	12	22	5	9	20
24	4	21	11	14	13	18	5	4	8	22	23
25	2	14	6	26	12	8	4	17	23	19	7
26	3	6	14	25	9	15	4	18	20	10	13

#1 ORBITS: (1,25)(2,17,14,11)(3,26,19,10)(4,8,7,6)(5,18,15,9)(13,21)(16,24)(22,23)
NUMBER OF DIFFERENT NEIGHBOUR TYPES = 6

#2 ORBITS: (1,7,4)(2,25,11)(3,23,10)(5,26,21,19,13,9)(6,24,18,16,15,8)(14,17)
NUMBER OF DIFFERENT NEIGHBOUR TYPES = 6

$n = 26,\ n_1 = 10,\ \#3$

VER	TYPE	NEIGHBOURS									
1	1	2	3	4	5	6	7	8	9	10	11
2	1	1	3	4	5	12	13	14	15	17	16
3	1	1	2	4	6	12	18	21	22	20	19
4	3	1	2	3	7	13	18	26	23	24	25
5	8	1	2	8	9	14	15	20	24	19	23
6	8	1	3	8	9	22	21	25	17	23	16
7	3	1	10	11	4	16	19	23	26	21	14
8	4	26	13	22	14	20	6	5	25	10	1
9	4	26	15	18	21	5	17	24	6	1	11
10	9	1	7	11	8	16	12	20	25	15	18
11	9	1	7	10	9	19	12	17	24	22	13
12	8	2	3	14	15	21	22	25	10	24	11
13	3	2	16	17	4	20	11	24	26	8	22
14	4	26	7	21	8	19	12	5	25	17	2
15	4	26	9	18	22	5	10	23	12	2	16
16	9	2	13	17	15	20	6	10	23	21	7
17	9	2	13	16	14	11	6	19	25	9	18
18	3	3	19	20	4	17	10	25	26	9	15
19	9	3	18	20	22	17	5	11	23	14	7
20	9	3	18	19	21	10	5	16	24	8	13
21	4	26	7	14	9	16	12	6	24	20	3
22	4	26	8	13	15	6	11	23	12	3	19
23	4	4	24	25	7	5	6	19	16	15	22
24	4	4	23	25	13	5	12	20	11	9	21
25	4	4	23	24	18	6	12	17	10	8	14
26	2	7	14	21	4	8	9	13	18	22	15

$n = 26,\ n_1 = 10,\ \#4$

VER	TYPE	NEIGHBOURS									
1	3	2	3	4	5	6	7	8	9	10	11
2	4	1	3	4	5	12	13	15	14	17	16
3	4	1	2	4	6	12	18	21	19	22	20
4	4	1	2	3	7	13	18	25	23	26	24
5	7	1	8	9	2	20	24	14	15	23	19
6	7	1	8	10	3	16	26	19	21	25	14
7	7	1	8	11	4	17	22	23	25	21	15
8	6	1	5	6	7	20	24	16	26	22	17
9	4	1	10	11	5	12	13	24	20	21	25
10	4	1	9	11	6	12	18	26	16	15	23
11	4	1	9	10	7	13	18	22	17	14	19
12	5	2	3	15	17	22	21	10	24	26	9
13	5	2	4	14	16	26	25	11	20	22	9
14	10	2	13	17	5	11	26	21	19	23	6
15	10	2	12	16	5	10	22	25	23	19	7
16	11	15	25	2	10	13	6	17	18	20	8
17	11	14	21	2	11	12	7	16	18	24	8
18	5	3	4	19	20	24	23	11	16	17	10
19	10	3	18	22	6	11	24	15	14	25	5
20	11	21	23	3	9	18	5	22	13	16	8
21	10	3	12	20	6	9	17	23	25	14	7
22	11	15	19	7	12	11	3	8	26	13	20
23	10	4	18	26	7	10	20	14	15	21	5
24	11	19	25	5	18	9	4	8	17	12	26
25	10	4	13	24	7	9	16	19	21	15	6
26	11	14	23	6	13	10	4	8	22	12	24

#3 ORBITS: (1,3,2)(5,12,6)(7,18,13)(9,21,22,14,15,8)(11,20,19,17,16,10)(23,25,24)
NUMBER OF DIFFERENT NEIGHBOUR TYPES = 6

#4 ORBITS: (2,4,3)(5,7,6)(9,11,10)(12,18,13)(14,25,23,21,19,15)(16,26,24,22,20,17)
NUMBER OF DIFFERENT NEIGHBOUR TYPES = 7

$n = 26,\ n_1 = 10,\ \#5$

VER	TYPE	NEIGHBOURS									
1	3	2	3	4	5	6	7	8	9	10	11
2	4	4	1	3	13	5	12	15	17	14	16
3	4	4	1	2	18	6	12	21	22	19	20
4	4	1	2	3	7	13	18	25	23	24	26
5	7	2	1	15	14	8	9	19	23	24	20
6	7	3	1	21	19	8	10	14	25	26	16
7	7	1	8	11	4	17	22	23	25	20	16
8	10	5	15	24	1	22	21	26	7	6	17
9	9	1	10	11	5	12	18	20	15	26	25
10	9	1	9	11	6	12	13	16	21	24	23
11	4	1	9	10	7	18	13	22	17	14	19
12	5	2	3	16	17	22	20	10	26	24	9
13	5	2	15	4	17	25	21	24	11	19	10
14	10	5	2	23	19	16	17	18	6	11	26
15	10	2	13	16	5	25	21	22	9	8	18
16	11	15	22	2	25	12	7	14	6	10	23
17	11	2	14	12	13	26	11	20	21	8	7
18	5	3	21	4	22	23	15	26	11	14	9
19	10	6	3	25	14	20	22	13	5	11	24
20	11	17	21	7	12	23	3	25	9	5	19
21	10	3	18	20	6	23	15	17	10	8	13
22	11	3	19	12	18	24	11	16	15	8	7
23	6	4	7	18	24	16	20	14	21	10	5
24	11	8	22	5	26	19	12	23	4	13	10
25	6	4	7	13	26	16	20	15	19	6	9
26	11	8	17	6	24	14	12	25	4	18	9

$n = 26,\ n_1 = 10,\ \#6$

VER	TYPE	NEIGHBOURS									
1	9	2	3	4	5	6	7	8	9	10	11
2	4	3	1	4	12	5	13	15	16	14	17
3	4	4	1	2	18	6	12	19	22	21	20
4	4	2	1	3	13	7	18	25	24	23	26
5	5	1	2	8	9	15	14	24	20	19	23
6	5	1	3	8	10	21	19	26	16	14	25
7	5	1	4	9	10	25	23	22	17	15	21
8	11	22	24	15	21	5	13	26	6	1	11
9	11	16	22	14	25	18	7	5	20	11	1
10	11	16	24	6	12	19	23	1	11	17	7
11	6	1	8	9	10	13	26	20	18	17	12
12	7	3	2	22	20	15	16	11	23	26	10
13	7	2	4	16	17	24	25	11	21	8	20
14	10	2	5	16	17	9	23	6	18	21	26
15	11	17	19	2	7	5	25	12	22	8	26
16	10	12	10	22	2	24	6	9	13	14	25
17	11	10	19	7	11	15	18	21	13	2	14
18	7	4	3	24	26	19	22	11	14	17	9
19	11	20	25	3	5	6	15	18	24	10	17
20	11	9	25	5	11	19	13	23	12	3	21
21	10	3	6	22	20	8	14	7	13	23	17
22	10	15	7	8	12	21	9	24	3	16	18
23	10	4	7	24	26	10	21	5	12	14	20
24	10	13	8	16	4	22	5	10	18	23	19
25	11	15	26	7	19	4	6	9	20	13	16
26	11	8	15	6	11	25	12	14	18	4	23

#5 { ORBITS: (2,3)(5,6)(9,10)(13,18)(14,19)(15,21)(16,20)(17,22)(23,25)(24,26)
NUMBER OF DIFFERENT NEIGHBOUR TYPES = 8

#6 { ORBITS: (2,3,4)(5,6,7)(9,10,8)(13,18,12)(15,19,25)(14,21,23)(17,20,26)(16,22,24)
NUMBER OF DIFFERENT NEIGHBOUR TYPES = 7

$n = 26,\ n_1 = 10$, #7

VER	TYPE	NEIGHBOURS									
1	2	2	3	4	5	6	7	8	9	10	11
2	9	1	3	4	5	12	13	15	14	17	16
3	2	2	1	4	12	6	18	21	22	19	20
4	2	2	1	3	13	7	18	25	24	23	26
5	9	19	20	23	24	14	15	9	8	2	1
6	2	8	1	10	25	3	26	21	14	19	16
7	2	9	1	11	22	4	16	25	17	23	19
8	9	1	6	10	5	25	17	15	24	22	18
9	9	1	7	11	5	22	21	14	24	12	26
10	2	8	1	6	17	11	26	13	12	20	23
11	2	9	1	7	21	10	16	13	15	20	18
12	9	10	23	26	17	15	9	2	22	21	3
13	9	10	11	20	17	21	14	24	2	25	4
14	9	6	16	26	25	2	9	13	22	5	20
15	9	11	16	18	21	2	8	12	25	5	23
16	2	14	6	26	2	19	18	17	15	7	11
17	9	7	16	19	22	2	24	12	8	13	10
18	2	15	11	16	8	20	26	22	24	3	4
19	2	5	20	23	24	3	7	21	17	6	16
20	2	5	19	23	14	3	10	22	13	18	11
21	9	3	6	19	12	25	24	15	9	13	11
22	9	3	18	20	12	8	14	17	9	25	7
23	2	5	19	20	15	7	10	25	12	4	26
24	9	4	18	26	13	8	9	17	21	5	19
25	9	4	7	23	13	22	15	14	21	8	6
26	2	12	10	23	9	6	4	14	24	16	18

$n = 26,\ n_1 = 10$, #8

VER	TYPE	NEIGHBOURS									
1	3	2	3	4	5	6	7	8	9	10	11
2	3	1	3	4	5	12	13	15	14	17	16
3	3	1	2	4	6	12	18	22	21	19	20
4	3	1	2	3	7	13	18	26	24	25	23
5	5	1	8	2	9	15	20	14	24	23	19
6	5	1	8	3	10	22	25	21	14	16	23
7	5	1	8	4	11	26	17	24	21	19	16
8	3	15	22	26	5	6	7	1	20	25	17
9	3	1	10	11	5	13	18	20	24	16	22
10	3	1	9	11	6	13	12	25	14	26	19
11	3	1	9	10	7	18	12	17	21	23	15
12	5	2	15	3	17	22	11	19	25	24	10
13	5	2	15	4	16	26	21	25	9	10	20
14	3	2	16	17	5	6	18	23	19	10	26
15	3	8	22	26	5	12	13	2	23	11	21
16	3	2	14	17	13	6	7	21	9	22	24
17	3	2	14	16	12	18	7	11	25	20	8
18	5	3	22	4	20	26	9	23	17	14	11
19	3	3	20	21	12	5	7	24	10	14	26
20	3	3	19	21	18	5	13	9	17	8	25
21	3	3	19	20	6	7	13	16	23	11	15
22	3	8	15	26	6	12	18	3	16	24	9
23	3	4	24	25	18	5	6	14	11	15	21
24	3	4	23	25	7	5	12	19	16	9	22
25	3	4	23	24	13	6	12	10	20	8	17
26	3	8	15	22	7	13	18	4	19	10	14

#7 ORBITS: (1,26,23,20,19,18,16,11,10,7,6,4,3)(2,25,24,22,21,17,15,14,13,12,9,8,5)
NUMBER OF DIFFERENT NEIGHBOUR TYPES = 2

#8 ORBITS: (1,26,25,24,23,22,21,20,19,17,16,15,14,11,10,9,8,4,3,2)(5,18,13,12,7,6)
NUMBER OF DIFFERENT NEIGHBOUR TYPES = 2

$n = 26,\ n_1 = 10$, #9

VER	TYPE	NEIGHBOURS									
1	7	2	3	4	5	6	7	8	9	10	11
2	11	5	12	1	16	3	17	4	15	13	14
3	10	2	12	13	1	19	21	20	6	7	18
4	10	2	14	15	1	22	25	24	7	6	23
5	10	1	2	8	9	12	16	22	23	18	26
6	11	10	24	1	26	4	16	3	19	25	18
7	11	22	26	4	21	23	17	1	3	20	11
8	10	5	12	22	1	19	24	14	11	10	20
9	10	5	18	23	1	25	21	13	11	10	15
10	10	1	6	8	9	24	26	14	21	13	17
11	11	7	20	1	17	8	16	9	25	19	15
12	10	3	2	21	19	17	5	24	25	8	23
13	11	2	3	14	15	20	19	10	9	23	26
14	5	2	17	4	13	25	10	22	20	18	8
15	11	4	24	2	22	16	21	13	19	11	9
16	5	2	17	5	15	26	11	18	24	6	20
17	11	7	26	11	21	16	10	25	12	2	14
18	10	6	3	16	25	20	21	5	14	9	22
19	5	3	6	12	13	25	26	8	15	11	22
20	11	3	13	7	18	23	14	11	16	24	8
21	5	3	7	12	18	17	22	24	9	10	15
22	11	5	8	18	26	14	19	21	7	4	15
23	5	4	24	7	25	20	12	26	9	13	5
24	11	8	12	10	20	21	23	6	16	15	4
25	10	4	6	14	23	18	19	17	9	12	11
26	11	6	10	16	19	17	13	5	22	7	23

$n = 26,\ n_1 = 10$, #10

VER	TYPE	NEIGHBOURS									
1	7	2	3	4	5	6	7	8	9	10	11
2	11	5	12	1	16	3	17	4	14	13	15
3	10	2	12	13	1	19	21	20	6	7	18
4	10	2	14	15	1	22	25	24	7	6	23
5	10	1	2	8	9	12	16	22	23	18	26
6	10	1	4	10	3	24	25	26	18	19	16
7	11	22	26	4	21	23	17	1	3	20	11
8	10	5	12	22	1	19	24	14	11	10	20
9	6	1	5	10	11	18	23	21	13	15	25
10	11	6	24	1	26	8	17	9	13	14	21
11	11	7	20	1	17	8	16	9	15	19	25
12	11	3	21	2	19	17	25	5	8	24	23
13	11	2	3	14	15	20	19	10	9	23	26
14	5	2	16	4	13	25	20	22	10	21	8
15	11	4	24	2	22	17	18	13	19	11	9
16	11	5	26	2	18	17	6	14	20	11	25
17	5	2	15	12	16	24	11	21	26	10	7
18	7	3	6	20	21	16	24	9	22	5	15
19	5	3	6	12	13	25	26	8	15	11	22
20	10	3	7	13	18	23	11	14	24	16	8
21	11	3	7	12	18	17	22	25	9	10	14
22	10	4	7	14	15	21	26	8	18	19	5
23	5	4	24	7	25	20	12	26	9	13	5
24	11	8	12	10	20	17	23	6	18	15	4
25	11	6	16	4	19	14	11	23	12	21	9
26	11	6	10	16	19	17	13	5	22	7	23

#9 { ORBITS: 26 POINTS
NUMBER OF DIFFERENT NEIGHBOUR TYPES = 4

#10 { ORBITS: (1,18)(10,21)(8,22)(6,3)(11,15)(2,16)(7,24)(4,20)(13,25)(26,12)
NUNBER OF DIFFERENT NEIGHBOUR TYPES = 5

NEIGHBOUR GRAPHS OF 26-FAMILY

(#1)

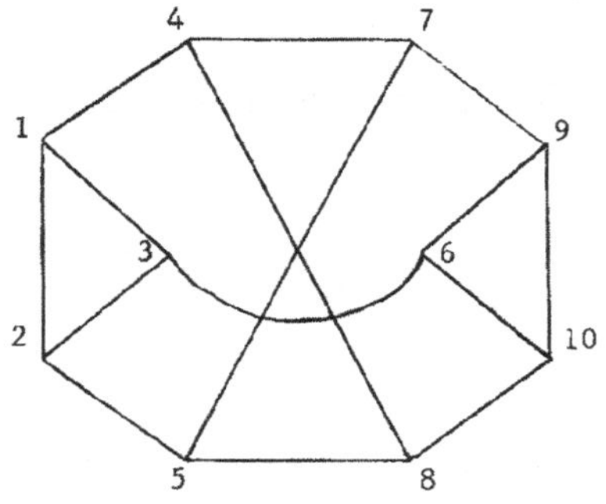

(#2)

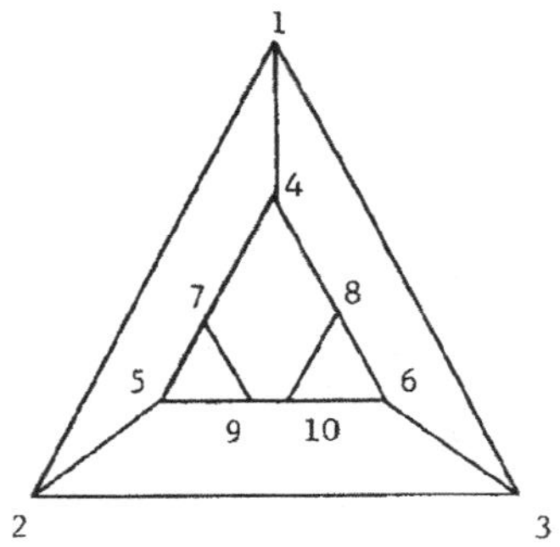

(#3)

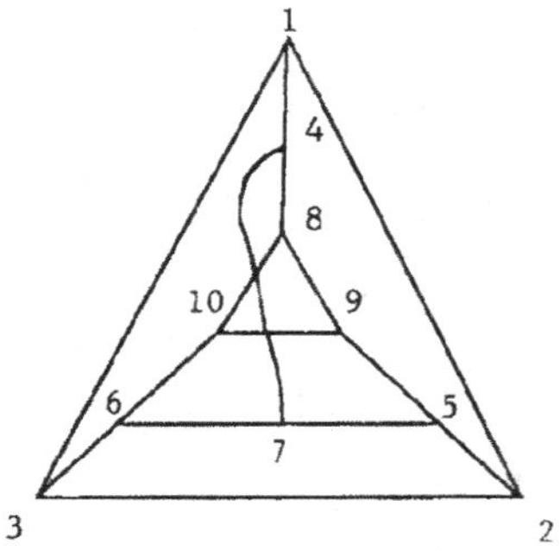

#4)

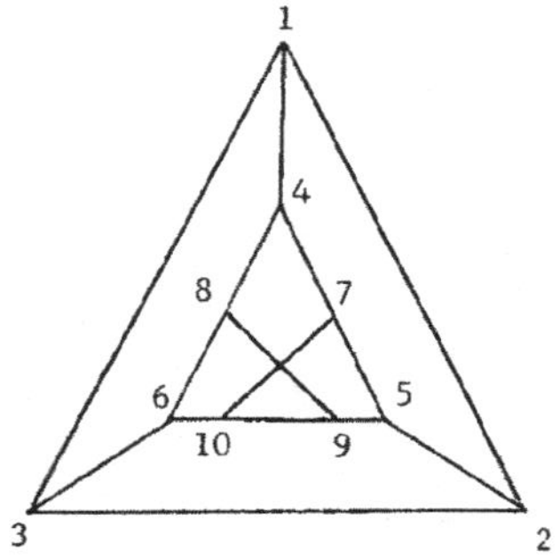

(#5)

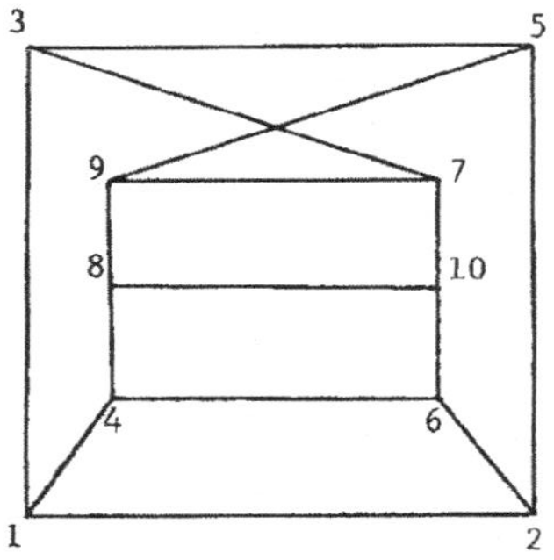

(#6)

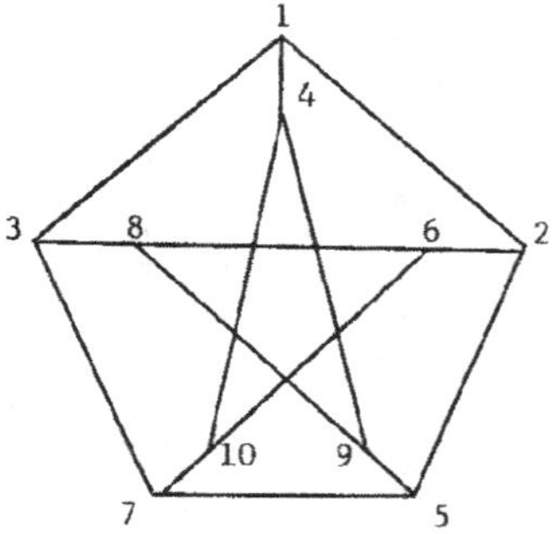

(#7)

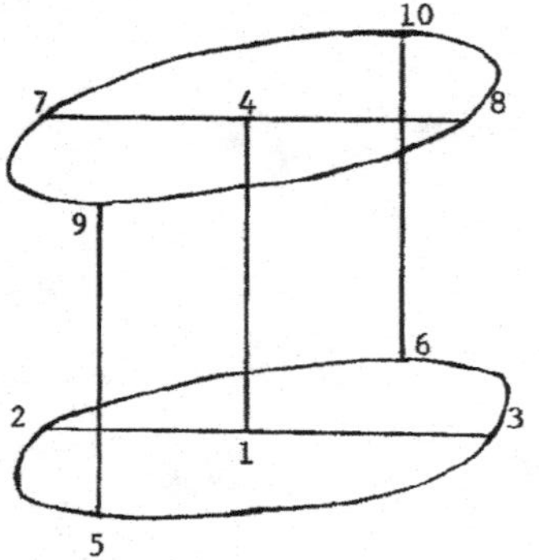
10
7
4
8
9
6
2
3
1
5

(#8)

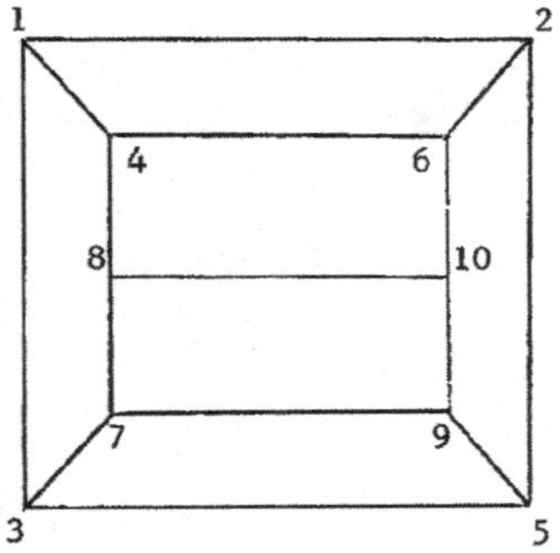
1
2
4
6
8
10
7
9
3
5

(#9)

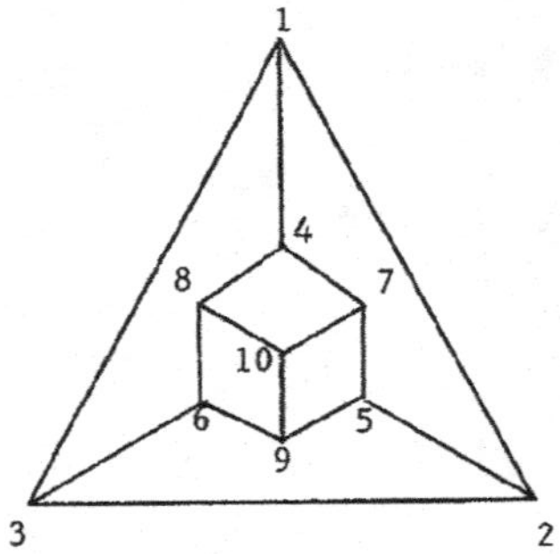
1
4
8
7
10
6
5
9
3
2

(#10)

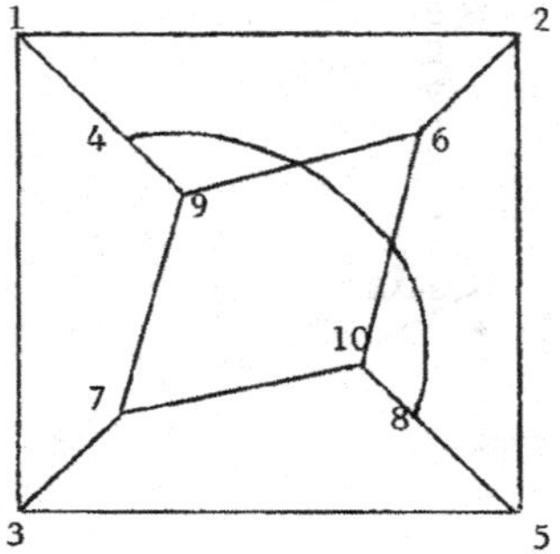
1
2
4
6
9
10
7
8
3
5

(#11)

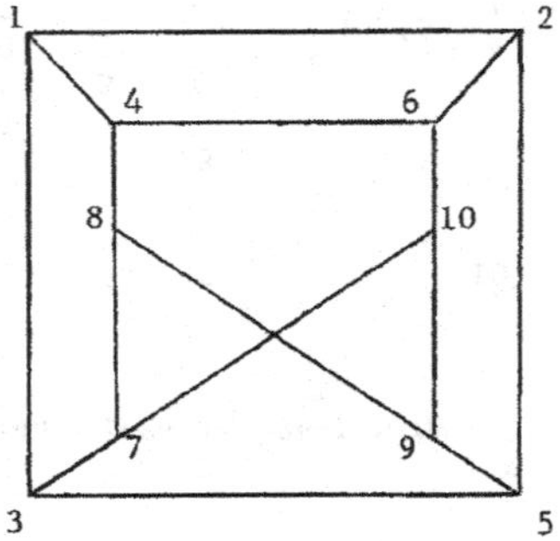
1
2
4
6
8
10
7
9
3
5

MULTIPLICITY OF THE NEIGHBOUR TYPES IN 26-FAMILY

TYPE \ #	1	2	3	4	5	6	7	8	9	10
1	3	3	3							
2	4	6	1				13			
3	6	9	4	1	1			20		
4	10	6	9	6	4	3				
5	1			3	3	3		6	5	4
6	2	1		1	2	1				1
7		1		3	3	3			1	2
8			3							
9			6		2	1	13			
10				6	5	6			9	7
11				6	6	9			11	12

$n = 27, n_1 = 10$, #1

VER	TYPE	NEIGHBOURS									
1	1	2	3	4	5	6	7	8	9	10	11
2	1	1	3	12	19	13	18	14	17	15	16
3	1	1	2	20	27	21	26	22	25	23	24
4	1	1	5	12	23	13	22	14	21	15	20
5	1	1	4	16	27	17	26	18	25	19	24
6	1	1	7	12	25	13	24	16	21	17	20
7	1	1	6	14	27	15	26	18	23	19	22
8	1	1	9	12	26	14	24	16	22	18	20
9	1	1	8	13	27	15	25	17	23	19	21
10	1	1	11	12	27	15	24	17	22	18	21
11	1	1	10	13	26	14	25	16	23	19	20
12	1	2	19	4	23	6	25	8	26	10	27
13	1	2	18	4	22	6	24	9	27	11	26
14	1	2	17	4	21	7	27	8	24	11	25
15	1	2	16	4	20	7	26	9	25	10	24
16	1	2	15	5	27	6	21	8	22	11	23
17	1	2	14	5	26	6	20	9	23	10	22
18	1	2	13	5	25	7	23	8	20	10	21
19	1	2	12	5	24	7	22	9	21	11	20
20	1	3	27	4	15	6	17	8	18	11	19
21	1	3	26	4	14	6	16	9	19	10	18
22	1	3	25	4	13	7	19	8	16	10	17
23	1	3	24	4	12	7	18	9	17	11	16
24	1	3	23	5	19	6	13	8	14	10	15
25	1	3	22	5	18	6	12	9	15	11	14
26	1	3	21	5	17	7	15	8	12	11	13
27	1	3	20	5	16	7	14	9	13	10	12

ORBITS: TRANSITIVE

NUMBER OF DIFFERENT NEIGHBOUR TYPES = 1

NEIGHBOUR GRAPH

$n = 28,\ n_1 = 12$, #1

VER	TYPE	NEIGHBOURS											
1	1	2	3	4	5	6	7	8	9	10	11	12	13
2	2	1	3	4	5	6	7	8	14	15	16	17	18
3	2	1	2	5	4	7	6	9	14	19	20	22	21
4	2	1	2	6	3	8	5	10	15	20	19	24	23
5	2	1	3	7	2	9	4	11	19	16	15	25	23
6	2	1	4	8	2	10	3	12	20	17	14	26	21
7	2	1	5	9	3	11	2	13	16	22	14	27	18
8	2	1	6	10	4	12	2	13	17	24	15	28	18
9	2	1	7	11	5	13	3	12	22	25	19	28	21
10	2	1	8	12	6	13	4	11	24	26	20	27	23
11	2	1	9	13	7	12	5	10	25	27	16	26	23
12	2	1	10	13	8	11	6	9	26	28	17	25	21
13	2	1	11	12	9	10	7	8	27	28	22	24	18
14	2	18	7	16	27	2	22	17	3	26	20	6	21
15	2	18	8	17	28	2	24	16	4	25	19	5	23
16	2	18	2	14	17	7	15	27	5	26	25	11	23
17	2	18	2	15	16	8	14	28	6	25	26	12	21
18	1	2	16	17	14	15	7	8	27	28	22	24	13
19	2	21	9	22	28	3	25	20	5	24	15	4	23
20	2	21	3	14	22	6	19	26	4	27	24	10	23
21	1	3	20	22	14	19	6	9	26	28	17	25	12
22	2	18	13	27	24	7	28	14	9	20	19	3	21
23	1	4	19	24	15	20	5	10	25	27	16	26	11
24	2	18	13	28	22	8	27	15	10	19	20	4	23
25	2	21	12	28	17	9	26	19	11	15	16	5	23
26	2	21	6	17	14	12	20	25	10	16	27	11	23
27	2	18	7	22	14	13	16	24	11	20	26	10	23
28	2	18	8	24	15	13	17	22	12	19	25	9	21

$n = 28,\ n_1 = 12$, #2

VER	TYPE	NEIGHBOURS											
1	3	2	3	4	5	6	7	8	9	10	11	12	13
2	3	1	3	4	5	6	7	8	14	15	16	17	18
3	3	1	2	4	5	6	7	8	19	20	21	22	23
4	4	1	2	3	5	6	9	10	14	15	19	20	24
5	4	1	2	3	4	7	9	11	14	16	19	21	25
6	4	1	2	3	4	8	10	12	15	17	20	22	26
7	4	1	2	3	5	8	11	13	16	18	21	23	27
8	4	1	2	3	6	7	12	13	17	18	22	23	28
9	3	14	19	4	5	24	25	28	1	10	11	13	12
10	3	15	20	4	6	24	26	27	1	9	12	13	11
11	3	16	21	5	7	25	27	26	1	9	13	12	10
12	3	17	22	6	8	26	28	25	1	10	13	11	9
13	3	18	23	7	8	27	28	24	1	11	12	10	9
14	3	9	19	4	5	24	25	28	2	15	16	18	17
15	3	10	20	4	6	24	26	27	2	14	17	18	16
16	3	11	21	5	7	25	27	26	2	14	18	17	15
17	3	12	22	6	8	26	28	25	2	15	18	16	14
18	3	13	23	7	8	27	28	24	2	16	17	15	14
19	3	9	14	4	5	24	25	28	3	20	21	23	22
20	3	10	15	4	6	24	26	27	3	19	22	23	21
21	3	11	16	5	7	25	27	26	3	19	23	22	20
22	3	12	17	6	8	26	28	25	3	20	23	21	19
23	3	13	18	7	8	27	28	24	3	21	22	20	19
24	4	9	14	19	4	28	10	13	15	18	20	23	27
25	4	9	14	19	5	28	11	12	16	17	21	22	26
26	4	10	15	20	6	27	12	11	17	16	22	21	25
27	4	10	15	20	24	26	13	11	18	16	23	21	7
28	4	9	14	19	24	25	13	12	18	17	23	22	8

#1 ORBITS: (2,25,27,26,11,28,24,13,12,10,19,22,20,9,16,15,17,8,14,5,7,4,6,3)(1,23,21,18)
NUMBER OF DIFFERENT NEIGHBOUR TYPES = 2

#2 ORBITS: (1,23,18,21,16,22,17,20,15,19,14,3,2,13,11,12,10,9)(4,27,26,28,25,24,7,8,6,5)
NUMBER OF DIFFERENT NEIGHBOUR TYPES = 2

$n = 28,\ n_1 = 12$, #3

VER	TYPE	NEIGHBOURS											
1	5	2	3	4	5	6	7	8	9	10	11	12	13
2	5	1	3	4	5	6	7	8	14	15	16	17	18
3	5	4	1	2	5	6	7	19	9	14	20	21	22
4	5	3	1	2	5	6	7	19	10	15	23	24	25
5	5	6	1	2	3	4	8	19	11	16	20	23	26
6	5	5	1	2	3	4	8	19	12	17	21	24	27
7	6	1	3	4	2	9	10	13	14	22	15	25	18
8	6	1	5	6	2	11	12	13	16	26	17	27	18
9	5	10	1	13	11	12	7	28	3	22	20	21	14
10	5	9	1	13	11	12	7	28	4	25	23	24	15
11	5	12	1	13	9	10	8	28	5	26	20	23	16
12	5	11	1	13	9	10	8	28	6	27	21	24	17
13	5	1	9	10	11	12	7	8	22	25	26	27	18
14	5	9	3	22	20	21	7	28	2	18	16	17	15
15	5	10	4	25	23	24	7	28	2	18	16	17	14
16	5	11	5	26	20	23	8	28	2	18	14	15	17
17	5	12	6	27	21	24	8	28	2	18	14	15	16
18	5	2	14	15	16	17	7	8	22	25	26	27	13
19	6	3	5	6	4	20	21	22	23	26	24	27	25
20	5	21	3	22	9	14	19	28	5	26	11	16	23
21	5	20	3	22	9	14	19	28	6	27	12	17	24
22	5	3	9	14	20	21	7	19	13	18	26	27	25
23	5	20	5	26	11	16	19	28	4	25	10	15	24
24	5	21	6	27	12	17	19	28	4	25	10	15	23
25	5	4	10	15	23	24	7	19	13	18	26	27	22
26	5	5	11	16	20	23	8	19	13	18	22	25	27
27	5	6	12	17	21	24	8	19	13	18	22	25	26
28	6	9	11	12	10	20	21	14	23	16	24	17	15

$n = 28,\ n_1 = 12$, #4

VER	TYPE	NEIGHBOURS											
1	7	2	3	4	5	6	7	8	9	10	11	12	13
2	7	1	3	4	5	6	7	8	14	15	16	17	18
3	7	1	2	4	5	6	7	9	14	19	20	21	22
4	7	1	2	3	5	6	7	10	15	19	23	24	25
5	7	1	2	3	4	6	7	11	16	20	23	26	27
6	7	1	2	3	4	5	7	12	17	21	24	26	28
7	7	1	2	3	4	5	6	13	18	22	25	27	28
8	7	1	9	10	11	12	13	2	14	15	16	17	18
9	7	1	8	10	11	12	13	3	14	19	20	21	22
10	7	1	8	9	11	12	13	4	15	19	23	24	25
11	7	1	8	9	10	12	13	5	16	20	23	26	27
12	7	1	8	9	10	11	13	6	17	21	24	26	28
13	7	1	8	9	10	11	12	7	18	22	25	27	28
14	7	2	8	15	16	17	18	3	9	19	20	21	22
15	7	2	8	14	16	17	18	4	10	19	23	24	25
16	7	2	8	14	15	17	18	5	11	20	23	26	27
17	7	2	8	14	15	16	18	6	12	21	24	26	28
18	7	2	8	14	15	16	17	7	13	22	25	27	28
19	7	3	9	14	20	21	22	4	10	15	23	24	25
20	7	3	9	14	19	21	22	5	11	16	23	26	27
21	7	3	9	14	19	20	22	6	12	17	24	26	28
22	7	3	9	14	19	20	21	7	13	18	25	27	28
23	7	4	10	15	19	24	25	5	11	16	20	26	27
24	7	4	10	15	19	23	25	6	12	17	21	26	28
25	7	4	10	15	19	23	24	7	13	18	22	27	28
26	7	5	11	16	20	23	27	6	12	17	21	24	28
27	7	5	11	16	20	23	26	7	13	18	22	25	28
28	7	6	12	17	21	24	26	7	13	18	22	25	27

#3 { ORBITS: (1,25,24,23,15,27,26,18,17,16,22,21,20,14,10,13,12,11,9,4,6,5,2,3)(7,28,19,8)
NUMBER OF DIFFERENT NEIGHBOUR TYPES = 2

#4 { ORBITS: TRANSITIVE
NUMBER OF DIFFERENT NEIGHBOUR TYPES = 1

V. SOME PROPERTIES OF 25- AND 26- FAMILIES.

Below we expose some results of the computer-aided analysis of the graphs of the 25- and 26-families. The numeration is that of the preceeding Section.

Results of this Section partially overlap with results of [Sh 4], [Sh 3], [Sh 5], [Pa 1], [Se 5]. Our results were announced in [Ro 1], [Ar 2].

1. 26-family and Steiner triple systems on 13 points.

There exist (cf., [Ha 3]) 2 non-isomorphic Steiner triple systems on 13 points. The corresponding graphs (whose vertices are triples) are # 7 (corresponding to the cyclic Steiner triple system) and # 3 in 26-family.

In [Sh 4] the authors took two non-coinciding representations of the cyclic Steiner triple system and derived from them 5 graphs (# 1, # 2, # 6, # 7, # 8) of the 26-family and 7 graphs (by descent, cf., 3 below) of the 25-family. The table in subsection 3 below shows that if the authors of the cited paper had not been so unlucky, they could have found using descent-ascent all the graphs of the 25- and 26-families. However,it is not clear how they would be able to establish that they found all graphs with these parameters.

2. Complement in the 25-family.

If A is the adjacency matrix of a strongly regular graph belonging to the 25-family then

$$\overline{A} = \tilde{I}_{25} - A$$

is also one. Below the number of the class of isomorphism of $\overline{A}$ is given as a function of the number of A

# of A	1	2	3	4	5	6	7	8	9	10	11	12	13	14	15
# of $\overline{A}$	13	7	4	3	8	9	2	5	6	12	14	10	1	11	15

3. Descent from the 26-family to the 25-family.

Let Γ be a strongly regular graph with 26 vertices, $x \in V(\Gamma)$. Let V_1

(resp. V_2) be the set of vertices of Γ which are incident (resp. non-incident) to x. Let Γ_x be the graph obtained in the following manner:

a) $V(\Gamma_x) = V(\Gamma) - x$;

b) the vertices of V_1 (resp., of V_2) are incident in Γ_x if and only if they are incident in Γ;

c) the vertices of V_1 are incident in Γ_x to vertices of V_2 if and only if they were not incident in Γ.

It is easily checked that Γ_x is strongly regular and belongs to the 25-family (cf., e.g., [Sh 4]).

In the table below in the position (m, n) stands the multiplicity of the m-th graph of the 25-family as a graph Γ_x of the n-th graph of the 26-family.

	1	2	3	4	5	6	7	8	9	10
1			3	3	3				3	3
2			3	3	3				3	3
3			6	6	6				6	6
4			6	6	6				6	6
5						13	13			
6			1	1	1				1	1
7			3	3	3				3	3
8						13	13			
9			1	1	1				1	1
10	12	12								
11	1	1								
12	12	12								
13			3	3	3				3	3
14	1	1								
15								26		

3.1. Remark. The graphs of the 26-family split into 4 groups: {1, 2}, {3, 4, 5, 9, 10}, {6, 7}, {8} and the columns of the above table are the same within one group. The 25-family splits accordingly into groups: {10, 11, 12, 14}, {1, 2, 3, 4, 6, 7, 9, 13}, {5, 8}, {15}.

It is interesting to compare this partition with the tables of the multiplicity of the neighbor types in the corresponding families and also with tables 4.2.1, 4.3.1.

4. Coinciding rows.

4.1. The matrices A constructed by the algorithm of Section T may be very close to each other. The i-th column of the table below contains the number of the matrices A, constructed by the algorithm of Section T, such that A and the matrix B constructed next to A have i coinciding rows (i.e., for i values of q one has $s_q(A) = s_q(B)$, where $s_q(C)$ is q-th row of C). The first row of the table shows i, the rows marked 25, 26 correspond to the 25- and 26-family.

NUMBER OF CONSECUTIVE GRAPHS WITH i COINCIDING ROWS

i	0	1	2	3	4	5	6	7	8	9	10	11
25	0	0	3	20	300	900	900	1500	750	700	900	500
26	0	0	1	5	25	100	350	280	450	260	300	400

Table continued

i	12	13	14	15	16	17	18	19	20
25	250	0	0	1750	0	1550	0	0	0
26	70	120	0	0	800	0	68	0	0

4.2. The 26-family.

In 4.1 we pointed out that the matrices successively constructed by the algorithm may have many common rows. The two tables below point out which isomorphism classes are close in that sense. The number "1" which stands at the intersection of the i-th column and the j-th row of Table 4.2.1 (resp. 4.2.2) indicates that among the matrices constructed by the algorithm there is a pair of successive ones which has 18 (resp. 16) common rows and such that the first matrix of the pair belongs to the i-th isomorphism class and the second one to the j-th one.

4.2.1. Table (18 common rows in the 26-family).

	1	2	3	4	5	6	7	8	9	10
1		1								
2	1									
3										
4					1					1
5				1	1	1			1	1
6					1				1	
7										
8										
9					1	1			1	1
10				1	1				1	

This induces the partition {1,2}, {3}, {4,5,6,9,10}, {7}, {8} of the graphs of the 26-family. Compare with Remark 3.1 above and with the remark after Table 4.3.1 below.

4.2.2. Table (16 common rows in the 26-family).

	1	2	3	4	5	6	7	8	9	10
1	1	1								
2	1									
3			1		1					
4					1	1				1
5			1	1	1	1			1	1
6				1	1				1	
7										
8										
9					1	1			1	1
10				1	1				1	

4.2.3. If the $n \times n$-matrices A and B have m common rows, let us denote by A_B and B_A the $(n-m) \times (n-m)$-matrices obtained from A and B by deleting m common rows and columns. For the 26-family and for $m = 18$ or 16, it turns out that A_B and B_A are (adjacency) matrices of isomorphic graphs.

Therefore the operation of transition from A to B can be described in the following manner:

Remove from $\Gamma(A)$ some subgraph Γ_1 spanned by $(n-m)$ vertices, and replace it by an isomorphic one. Call this operation <u>surgery</u>.

It turns out that for $m = 18$ graphs A_B and B_A always belong to the class of isomorphism of the graph given below:

$$\begin{matrix} 0&1&1&1&1&0&0&0\\ 1&0&1&0&0&1&0&0\\ 1&1&0&0&0&1&0&0\\ 1&0&0&0&0&1&1&1\\ 1&0&0&0&0&1&1&1\\ 0&1&1&1&1&0&0&0\\ 0&0&0&1&1&0&0&1\\ 0&0&0&1&1&0&1&0 \end{matrix}$$

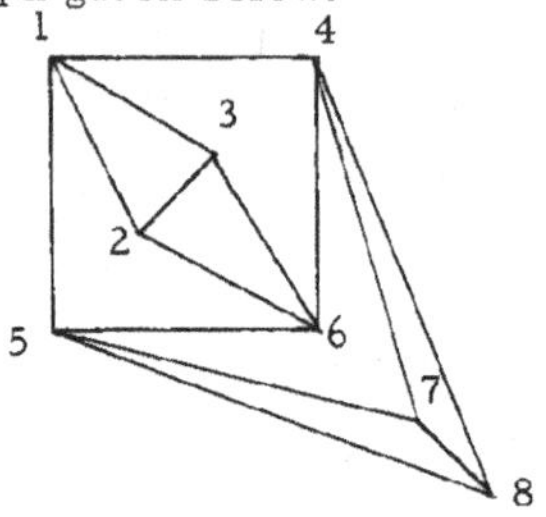

A program was written which found all (up to isomorphism) strongly regular completions of this graph to strongly regular graphs with 26 vertices. All completions belong to the isomorphism classes 1, 2, 4, 5, 6, 9, 10. This gives rise to the hypothesis that only these isomorphism classes have representatives which have 18 common rows. This hypothesis was then checked, and it turned out that it is true.

When $m = 16$ by random search four isomorphism classes of matrices A_B were found. We do not give them here.

4.3. <u>The 25-family</u>.

The Tables 4.3.1 and 4.3.2, given below have the same significance and are arranged analogously to Tables 4.2.1, 4.2.2.

4.3.1. <u>Table</u> (17 common rows in the 25-family).

	1	2	3	4	5	6	7	8	9	10	11	12	13	14	15
1				1				1							
2			1	1	1		1								
3		1		1				1					1		
4	1	1	1	1	1	1	1	1		1		1			
5		1		1									1		
6				1					1						
7		1		1				1							
8	1		1	1			1								
9						1									
10				1								1			

(Table continued)

	1	2	3	4	5	6	7	8	9	10	1	12	13	14	15
11															
12				1						1					
13			1		1										
14															
15															

This table induces the partition $\{1,2,3,4,5,6,7,8,9,10,12,13\}$, $\{11\}$, $\{14\}$, $\{15\}$. Compare with Remark 3.1 and with the remark after the Table 4.2.1.

4.3.2. Table (15 common rows in the 25-family).

	1	2	3	4	5	6	7	8	9	10	11	12	13	14	15
1	1			1				1							
2			1	1	1	1	1								
3		1		1	1		1	1					1		
4	1	1	1	1	1	1	1	1		1		1	1		
5		1	1	1									1		
6		1		1				1	1						
7		1	1	1				1	1						
8	1		1	1		1	1								
9				1	1	1	1								
10				1						1		1			
11															
12				1						1		1			
13			1	1	1							1			
14															
15															

4.3.3. Note that table 4.3.2 is not symmetric; this means that our material is insufficiently representative.

In all cases when (for $m = 15$ or 17) the matrices A_B and B_A were constructed, they proved to be isomorphic.

The matrices A_B, constructed for $m = 17$, are isomorphic to the matrix given in 4.2.3. Possibly this phenomenon is connected with the operation of descent.

5. Coinciding minors

Among the matrices constructed by the algorithm there were pairs A and B which had large common minors.

Systematic research has not yet been performed in that direction. However, some matrices of the 25-family have common minors of order 19. Such pairs are also contained among canonical forms given in the preceding Section. Below for matrices with numbers n, m from the preceding Section, those 19 vertices are pointed out which span coinciding minors.

# n	# m	numbers of vertices spanning (19 × 19) - minor
2	3	1 + 19
4	5	1 + 19
6	7	1 + 19
6	8	1 + 14, 16, 17, 18, 24, 25
6	9	1 + 14, 16, 17, 18, 22, 23
7	8	1 + 14, 16, 17, 18, 22, 23
7	9	1 + 14, 16, 17, 18, 20, 21
8	9	1 + 19.

6. More information on the 25- and 26-families can be found in [Pa 1, Se 5, Sh 4]. For example, the clique structure of these graphs and the classes of Seidel equivalence (switching) are found there.

Also note that the tables of the preceding Section are, in principle, sufficient to answer some questions, such as, what is the automorphism group of the graphs, etc.

AA. A GRAPHICAL REGULAR REPRESENTATION OF Sym(n).

The next three sections can be considered as examples of application of the stabilization procedures of Sections C and M. On the other hand, the questions discussed in Sections AA, AB, AC have attracted the attention of several authors (e.g., [Sa 1], [Wa 2], [Wa 3], [Im 1], [Im 2]).

1. Let G be a finite group. A graph Γ is said to be a graphical regular representation of G, if G acts simply transitively on vertices of Γ and $G \simeq \mathrm{Aut}\,\Gamma$.

Proposition. If Γ is a graph and $\mathfrak{A}(\Gamma) \simeq \mathbf{Z}[G]$ (isomorphism of cellular algebras) then Γ is a graphical regular representation of G.

Proof. By C 8.2 we have $\mathrm{Aut}\,\Gamma = \mathrm{Aut}\,\mathfrak{A}(\Gamma)$. Since $|V(\Gamma)| = |G|$, the only thing to prove is that $\mathrm{Aut}\,\Gamma \simeq G$. But it is well-known that $\mathrm{Aut}\,\mathbf{Z}[G] \simeq G$ (cf., [Ha 2]).

2. Let $\mathfrak{A} = \mathbf{Z}[G]$ be the group algebra of G (cf., G 1). The operators R_g, of the right multiplication by $g \in G$, form a standard basis of $\mathfrak{A}$. We shall identify g and R_g.

3. Below we use the stabilization procedure of Sections C and M to check that $\mathfrak{A}(\Gamma) \simeq Z[\mathrm{Sym}(n)]$ for an explicitly given graph Γ. Then the above proposition will give us the following

Theorem. There exists a simple graph Γ (without loops, multiple or directed edges) which has $n!$, $n > 3$, vertices, and whose automorphism group is isomorphic to $\mathrm{Sym}(n)$ and acts transitively on its vertices.

Proof. Set $p_1 = (1,2,3,4)$, $p_2 = (1,2,3)$, $p_3 = (1,4)$, $p_i = (i,i+1)$, $i \geq 4$,

$$\Gamma = p_1 + p_1^{-1} + p_2 + p_2^{-1} + \textstyle\sum_{i=3}^{n-1} p_i$$

According to our convention ($R_g \longleftrightarrow g$), this is an $n! \times n!$-matrix. Since the coefficients of the elements of G in Γ are 0 or 1, it is a (0,1)-matrix. Since Γ contains g^{-1} together with g, it is symmetric. So it remains (by Proposition 1) to show that

$$\mathfrak{A}(\Gamma) = \mathbf{Z}[G]$$

Consider Γ^2. One has

$$\Gamma^2 = (n-2)\cdot 1 + 2\cdot(1,3,2,4) + (1,2,3,4) + 2\cdot(1,4,2,3) + (1,4,3,2) + (1,2,4,3) + (1,3,4,2) + 2\cdot(1,2,3) + 2\cdot(1,3,2) + (2,3,4) + (2,4,3) + 2\cdot(1,3)(2,4) + 2\cdot(3,4) + 2\cdot(1,4) + (\text{terms, containing } 5,6,\ldots,n).$$

The summands of Γ which appear in Γ^2 are $p = p_2 + p_2^{-1} + p_3$ (with coefficient 2) and $b = p_1 + p_1^{-1}$ (with coefficient 1). By definition of the product ($X \circ X$, cf., C 4.2), we have $p, b, v = \sum_{i=4}^{n-1} p_i \in \mathfrak{A}(\Gamma)$.

Now consider p^2. One has

$$p^2 = 3\cdot 1 + (1,2,3) + (1,3,2) + (1,2,3,4) + (1,4,3,2) + (1,3,2,4) + (1,4,2,3).$$

Since $p_2 + p_2^{-1}$ and p_3 enter in p^2 with different coefficients and since $p^2 \in \mathfrak{A}(\Gamma)$, we have $p_2 + p_2^{-1}, p_3 \in \mathfrak{A}(\Gamma)$.

Now $p_3(p_2 + p_2^{-1}) = (1,4,2,3) + (1,4,3,2)$. Since of these two substitutions only $(1,4,3,2) = p_1^{-1}$ enters in Γ, we have $p_1^{-1} \in \mathfrak{A}(\Gamma)$, whence $p_1 \in \mathfrak{A}(\Gamma)$. It is easy to verify that p_1, p_3 generate Sym(n) acting on [1,2,3,4]. Therefore, $d = (3,4) \in \mathfrak{A}(\Gamma)$. Consider

$$d\left(\sum_{i=4}^{n-1} p_i\right) d = (3,5) + \sum_{i=5}^{n-1} p_i$$

It follows that $(3,5) \in \mathfrak{A}(\Gamma)$. Therefore, $(4,5) \in \mathfrak{A}(\Gamma)$.

Suppose now that $a = (q,q+1) \in \mathfrak{A}(\Gamma)$ and $v = \sum_{q+1}^{n-1} p_i \in \mathfrak{A}(\Gamma)$. Then

$$a\ v\ a = (q,q+2) + \sum_{q+2}^{n-1} p_i$$

This is the inductive step which concludes our proof. Namely, it shows that $\mathrm{Sym}(n) \subset \mathfrak{A}(\Gamma)$.

AB. A GRAPHICAL REGULAR REPRESENTATION OF $SL_n(\mathbb{F}_q)$.

1. Let $\mathbb{F}_q$ be the finite field with q elements, $q = p^m$, p a prime. Let H be the group of unimodular $(n + 1) \times (n + 1)$-matrices with entries in $\mathbb{F}_q$, $H = SL_{n+1}(\mathbb{F}_q)$.

<u>Theorem</u>. Suppose that $p > 5$. Then there exists a simple graph Γ such that $\mathrm{Aut}\,\Gamma \simeq H$, $|\Gamma| = |H|$.

<u>Remark</u>. This assertion and its proof can be generalized to rational points over $\mathbb{F}_q$ of semi-simple algebraic groups defined and split over $\mathbb{F}_q$ (cf., [Bo 1]).

2. To prove the theorem we shall construct, as in the preceding section a graph Γ such that $\mathfrak{A}(\Gamma) = \mathbb{Z}[H]$. Also, as in the preceding section, we shall write the elements of the group ring instead of the operators of right multiplication. Let us introduce some notations.

3. For

$$A = \begin{pmatrix} a & b \\ c & d \end{pmatrix} \in SL_2(\mathbb{F}_q)$$

set

$$\varphi_{ij}(A) = \begin{pmatrix}
1 & & & & & & & & & & \\
& 1 & & & & & & & & & \\
& & \ddots & & & & & & & & \\
& & & 1 & & & & & & & \\
& & & & \overset{i}{a} & & & \overset{j}{b} & & & \\
& & & & & 1 & & & & & \\
& & & & & & \ddots & & & & \\
& & & & & & & 1 & & & \\
& & & & c & & & & d & & \\
& & & & & & & & & 1 & \\
& & & & & & & & & & \ddots \\
& & & & & & & & & & & 1
\end{pmatrix} \quad \text{if } j > i$$

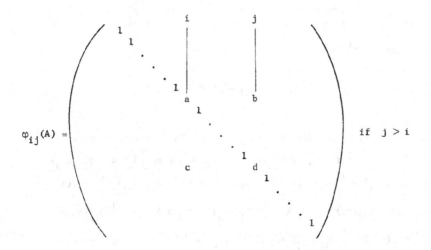

with zeros in all other places, and $\varphi_{ij}(A) = \varphi_{ji}(A^{t^{-1}})$ if $j < i$.

Set

$$n_a = \begin{pmatrix} 1 & a \\ 0 & 1 \end{pmatrix}, \; t = \begin{pmatrix} 0 & 1 \\ -1 & 0 \end{pmatrix}, \; d = \begin{pmatrix} 1 & -1 \\ 1 & 0 \end{pmatrix} = n_1 t^{-1}$$

$$v = t + t^{-1} + d + d^{-1} + n_1 + n_{-1}$$

Take $c \neq \pm 2$, a generator of the multiplicative group of $\mathbf{F}_q$, (there exists such c since $p > 5$) and set

$$u = n_c t^{-1} + t n_c^{-1}, \; \bar{v} = v + u$$

Next set for $n = 1$

$$w = \bar{w} = 0$$

and for $n \geq 2$

$$w = \varphi_{23}(t) \varphi_{34}(t) \ldots \varphi_{n\, n+1}(t)$$

$$\bar{w} = \varphi_{12}(n_1) w + w^{-1} \varphi_{12}(n_{-1}) + \varphi_{12}(n_4) w + w^{-1} \varphi_{12}(n_{-4})$$

Now set $m = v$ if $p = q$ and $m = \bar{v}$ if $p < q$. Also put $n = n_1 + n_4$, $n' = n_{-1} + n_{-4}$. Next set

$$\Gamma = \varphi_{12}(m) + \bar{w}$$

4. As in the preceding section, let us consider Γ^2. We have (for $n > 1$)

$$\begin{aligned}\Gamma^2 = {} & \varphi_{12}(m)^2 + 4 \cdot 1 + \varphi_{12}(n_1) w \varphi_{12}(n_1) w + \varphi_{12}(n_1) w \varphi_{12}(n_4) w + \varphi_{12}(n_{-3}) \\ & + w^{-1} \varphi_{12}(n_{-1}) w^{-1} \varphi_{12}(n_{-1}) + w^{-1} \varphi_{12}(n_3) w + w^{-1} \varphi_{12}(n_{-1}) w^{-1} \varphi_{12}(n_{-4}) \\ & + \varphi_{12}(n_4) w \varphi_{12}(n_1) w + \varphi_{12}(n_3) + \varphi_{12}(n_4) w \varphi_{12}(n_4) w + w^{-1} \varphi_{12}(n_{-3}) w \\ & + w^{-1} \varphi_{12}(n_{-4}) w^{-1} \varphi_{12}(n_{-1}) + w^{-1} \varphi_{12}(n_{-4}) w^{-1} \varphi_{12}(n_{-4}) + \varphi_{12}(m) \varphi_{12}(n_1) w\end{aligned}$$

$$+ \varphi_{12}(m)\, w^{-1}\, \varphi_{12}(n_{-1}) + \varphi_{12}(m)\, \varphi_{12}(n_4)\, w + \varphi_{12}(m)\, w^{-1}\, \varphi_{12}(n_{-4})$$

$$+ \varphi_{12}(n_1)\, w\, \varphi_{12}(m) + w^{-1}\, \varphi_{12}(n_{-1})\, \varphi(m) + \varphi_{12}(n_4)\, w\, \varphi_{12}(m)$$

$$+ w^{-1}\, \varphi_{12}(n_{-4})\, \varphi_{12}(m) =$$

$$= \varphi_{12}(m)^2 + 4 \cdot 1 + \varphi_{12}(n_1)\, \varphi_{13}(n_{-1})\, w^2 + \varphi_{12}(n_1)\, \varphi_{13}(n_{-4})\, w^2 + \varphi_{12}(n_{-3})$$

$$+ \varphi_{1n+1}(n_{-1})\, \varphi_{1n}(n_1)\, w^{-2} + \varphi_{1n+1}(n_3) + \varphi_{1n+1}(n_{-1})\, \varphi_{1n}(n_4)\, w^{-2}$$

$$+ \varphi_{12}(n_4)\, \varphi_{13}(n_{-1})\, w^2 + \varphi_{12}(n_3) + \varphi_{12}(n_4)\, \varphi_{13}(n_{-4})\, w^2 + \varphi_{1n+1}(n_{-3})$$

$$+ \varphi_{1n+1}(n_{-4})\, \varphi_{1n}(n_1)\, w^{-2} + \varphi_{1n+1}(n_{-4})\, \varphi_{1n}(n_4)\, w^{-2} + \varphi_{12}(mn_1)\, w$$

$$+ \varphi_{12}(m)\, \varphi_{1n+1}(n_{-1})\, w^{-1} + \varphi_{12}(mn_4)\, w + \varphi_{12}(m)\, \varphi_{1n+1}(n_{-4})\, w^{-1}$$

$$+ \varphi_{12}(n_1)\, \varphi_{13}(\widetilde{m})\, w + \varphi_{1n+1}(n_{-1}m)\, w^{-1} + \varphi_{12}(n_4)\, \varphi_{13}(\widetilde{m})\, w + \varphi_{1n+1}(n_{-4}m)\, w^{-1} =$$

(here $\widetilde{m}$ is defined by $w\, \varphi_{12}(m)\, w^{-1} = \varphi_{13}(\widetilde{m})$)

$$= 4 \cdot 1 + \varphi_{12}(m^2 + n_{-3} + n_3) + \varphi_{1n+1}(n_3 + n_{-3})\, \varphi_{12}(n_1 + n_4)\, \varphi_{13}(n_{-1} + n_{-4})\, w^2$$

$$+ \varphi_{1n+1}(n_{-1} + n_{-4})\, \varphi_{1n}(n_1 + n_4)\, w^{-2} + [\varphi_{12}(m(n_1 + n_4)) + \varphi_{12}(n_1 + n_4)\, \varphi_{13}(\widetilde{m})]\, w$$

$$+ [\varphi_{1n+1}((n_{-1} + n_{-4})m) + \varphi_{12}(m)\, \varphi_{1n+1}(n_{-1} + n_{-4})]\, w^{-1}.$$

For $n = 1$ we have $\Gamma^2 = \varphi_{12}(m^2)$.

We want to choose from this sum the terms which have coefficient ≥ 2.

It follows from the unicity properties of the Bruhat decomposition (cf., [Bo 1]) that only the following cases can occur:

a) $n = 1$, only summands of $\varphi_{12}(m^2)$ can have coefficient ≥ 2;

b) $n = 2$, only summands of $\varphi_{12}(m^2 + n_{-3} + n_3) + 2[\varphi_{12}(n_1 + n_4)\, \varphi_{13}(n_{-1} + n_{-4})]\, \varphi_{23}(-1)$ can have coefficient ≥ 2 (since $w^2 = w^{-2} = \varphi_{23}(-1)$ and $\varphi_{12}(n_a)\, \varphi_{13}(n_b) = \varphi_{13}(n_b)\, \varphi_{12}(n_a)$) in this case;

c) $n \geq 3$, only summands of $\varphi_{12}(m^2 + n_{-3} + n_3)$ can have coefficients ≥ 2.

Thus the following holds.

4.1. <u>Assertion</u>. The terms with coefficients ≥ 2 are contained in $\varphi_{12}(m^2)$ if $n = 1$;

$\varphi_{12}(m^2 + n_{-3} + n_3) + 2[\varphi_{12}(n_1 + n_4)\ \varphi_{13}(n_{-1} + n_{-4})]\ \varphi_{23}(-1)$ if $n = 2$;

$\varphi_{12}(m^2 + n_{-3} + n_3)$ if $n > 2$.

To compute $\varphi_{12}(m^2)$, put $a = 1 + n_1 + n_c$, $b = 1 + n_1$, $c = n_1 + n_{-1}$, $a' = 1 + n_{-1} + n_{-c}$, $b' = 1 + n_{-1}$

$$v = bt^{-1} + tb' + c$$

$$\bar{v} = at^{-1} + ta' + c$$

$$v^2 = bt^{-1}bt^{-1} + bb' + bt^{-1}c + tb'bt^{-1} + tb'tb' + tb'c + cbt^{-1} + ctb' + c^2.$$

$$\bar{v}^2 = at^{-1}at^{-1} + aa' + at^{-1}c + ta'at^{-1} + ta'ta' + ta'c + cat^{-1} + cta' + c^2.$$

Note first that $n_{\pm 3}$ does not enter in v^2 or $\bar{v}^2$, since $c \neq \pm 2$.

We are going next to use the uniqueness of the Bruhat decomposition in SL(2). To this end note that for $d \neq 0$, one has

$$\begin{pmatrix} 0 & 1 \\ -1 & 0 \end{pmatrix} \begin{pmatrix} 1 & d \\ 0 & 1 \end{pmatrix} \begin{pmatrix} 0 & 1 \\ -1 & 0 \end{pmatrix} = \begin{pmatrix} 1 & d^{-1} \\ 0 & 1 \end{pmatrix} \begin{pmatrix} 0 & -d^{-1} \\ d & 0 \end{pmatrix} \begin{pmatrix} 1 & d^{-1} \\ 0 & 1 \end{pmatrix}$$

It follows from this equality that we have

4.2 Assertion. The terms with coefficient 2 in $m^2 + n_3 + n_{-3}$ are all contained in v^2.

(Indeed if d above is not ± 1, then we get nontrivial diagonal element, which cannot be obtained in any other way.)

Let us now compute v^2. We have

$$v^2 = 6 \cdot 1 + t^2 + t^{-2} + n_1t^{-1}n_1t^{-1} + tn_{-1}tn_{-1} + n_2 + n_{-2} + tn_1t^{-1} + t^2n_{-1} + tn_1$$
$$+ tn_{-1} + t^{-1}n_1t^{-1} + n_{-1} + t^{-1}n_1 + t^{-1}n_{-1} + n_1 + n_1t^{-2} + n_1t^{-1}n_1 + n_1t^{-1}n_{-1}$$
$$+ tn_{-1}t + tn_{-1}t^{-1} + t + tn_{-2} + n_1t + n_1t^{-1} + n_2t^{-1} + n_1tn_{-1} + n_{-1}t + n_{-1}t^{-1}$$
$$+ t^{-1} + n_{-1}tn_{-1}.$$

Using the relations

$$tn_1t = n_1t^{-1}n_1, \quad tn_{-1}t = n_{-1}tn_{-1}$$

which we already used above, and also the equality $\begin{pmatrix} 0 & 1 \\ -1 & 0 \end{pmatrix} = -\begin{pmatrix} 0 & -1 \\ 1 & 0 \end{pmatrix}$, which connects t and t^{-1}, we get

$$v^2 = 6 \cdot 1 + (t^2 + t^{-2}) + n_2 t^{-1} n_1 + n_{-1} t n_{-2} + n_2 + n_{-2} + n_1 t n_1 + t^2 n_{-1} + t n_1 + t n_{-1}$$

$$+ n_1 t^{-1} n_1 + n_{-1} + t^{-1} n_1 + t^{-1} n_{-1} + n_1 + n_1 t^{-2} + n_1 t^{-1} n_1 + n_1 t^{-1} n_{-1} + n_{-1} t n_{-1}$$

$$+ n_{-1} t^{-1} n_{-1} + t + t n_{-2} + n_1 t + n_1 t^{-1} + n_2 t^{-1} + n_1 t n_{-1} + n_{-1} t + n_{-1} t^{-1} + t^{-1}$$

$$+ n_{-1} t n_{-1}.$$

It is seen from this expression that the coefficient 2 has only the following expression:

$$d = t^2 + n_1 t^{-1} n_1 + n_{-1} t n_{-1}$$

Therefore, it follows from 4.1 and from the computations above that the assertion below holds:

4.3. <u>Assertion</u>. i) If $n \neq 2$, then $f = \varphi_{12}(d) \in \mathfrak{a}(\Gamma)$.

ii) If $n = 2$, then $f = \varphi_{12}(d) + \varphi_{12}(n_1 + n_4)\, \varphi_{13}(n_{-1} + n_{-4})\, \varphi_{23}(-1) \in \mathfrak{a}(\Gamma)$.

We have

$$d^2 = 1 + n_1 t n_1 + n_{-1} t^{-1} n_{-1} + n_1 t n_1 + n_1 t^{-1} n_2 t^{-1} n_1 + 1 + n_{-1} t^{-1} n_{-1} + 1 + n_{-1} t n_{-2} t n_{-1}$$

$$= 3 \cdot 1 + 2[n_1 t n_1 + n_{-1} t^{-1} n_{-1}] + n_{-3} \begin{pmatrix} 0 & 2^{-1} \\ -2 & 0 \end{pmatrix} n_{-3} + n_3 \begin{pmatrix} 0 & -2^{-1} \\ 2 & 0 \end{pmatrix} n_3.$$

Further

$$\varphi_{12}(d)[\varphi_{12}(d) + \varphi_{12}(n_1 + n_4)\, \varphi_{13}(n_{-1} + n_{-4})\, \varphi_{23}(-1)]$$

$$= \varphi_{12}(d^2) + \varphi_{12}(dn_1 + dn_4)\, \varphi_{13}(n_{-1} + n_{-4})\, \varphi_{23}(-1) \quad \text{and}$$

$$[\varphi_{12}(d) + \varphi_{12}(n_1 + n_4)\, \varphi_{13}(n_{-1} + n_{-4})]\, \varphi_{23}(-1)\, \varphi_{12}(d)$$

$$= \varphi_{12}(d^2) + \varphi_{12}(n_1 t^2 + n_4 t^2)\, \varphi_{13}(n_1 + n_4)\, \varphi_{23}(-1) + \varphi_{12}(x)\, \varphi_{13}(y)\, \varphi_{23}(z) \quad \text{where}$$

$z \not\equiv \pm 1$, $z \neq 0$.

Next

$$\varphi_{12}(n_1 + n_4)\, \varphi_{13}(n_{-1} + n_{-4})\, \varphi_{23}(-1)\, \varphi_{12}(n_1 + n_4)\, \varphi_{13}(n_{-1} + n_{-4})\, \varphi_{23}(-1)$$

$$= \varphi_{12}((n_1 + n_4)(n_{-1} + n_{-4}))\, \varphi_{13}((n_{-1} + n_{-4})(n_1 + n_4))$$

$$= \varphi_{12}(2 \cdot 1 + n_{-3} + n_3)\, \varphi_{13}(2 \cdot 1 + n_{-3} + n_3)$$

$= 2 \cdot 1 + 2\varphi_{12}(n_3 + n_{-3}) + 2\varphi_{13}(n_3 + n_{-3}) + \varphi_{12}(n_3 + n_{-3})\ \varphi_{13}(n_3 + n_{-3})$.

It follows that

4.4. Assertion. The coefficient 2 in f^2 has

$g = \varphi_{12}(n_1 t n_1 + n_{-1} t^{-1} n_{-1})$ if $n \neq 2$

$g = \varphi_{12}(n_1 t n_1 + n_{-1} t^{-1} n_{-1} + n_3 + n_{-3}) + \varphi_{13}(n_3 + n_{-3})$ if $n = 2$.

Therefore $g \in \mathfrak{a}(\Gamma)$.

Now, if $n = 2$, then compare g with

$\varphi_{12}(m^2 + n_3 + n_{-3}) + 2[\varphi_{12}(n_1 + n_4)\ \varphi_{13}(n_{-1} + n_{-4})]\ \varphi_{23}(-1)$ (which belongs to $\mathfrak{a}(\Gamma)$ by 4.1). It follows that

$g' = \varphi_{12}(n_1 t n_1 + n_{-1} t^{-1} n_{-1} + n_3 + n_{-3}) \in \mathfrak{a}(\Gamma)$.

Now for $n \neq 2$ consider

$$s = fg \cap f \in \mathfrak{a}(\Gamma) ,$$

and for $n = 2$

$$s = fg' \cap f \in \mathfrak{a}(\Gamma).$$

We see that in both cases

$$s = \varphi_{12}(n_1 t^{-1} n_1 + n_{-1} t n_{-1})$$

which implies the following:

4.5. Assertion. $s = \varphi_{12}(n_1 t^{-1} n_1 + n_{-1} t n_{-1}) \in \mathfrak{a}(\Gamma)$ and

$\varphi_{12}(t^2) = \varphi_{12}(d) - s \in \mathfrak{a}(\Gamma)$.

Now consider $t^2 \cdot m + m$. The coefficient 2 in this expression has only

$$t + t^{-1}$$

whence

4.6. <u>Assertion</u>. $\varphi_{12}(t + t^{-1}) \in \mathfrak{a}(\Gamma)$.

Note that $\varphi_{12}(t + t^{-1}) \cdot s$ and Γ^2 have only one common term, namely $\varphi_{12}(n_{-1}tn_{-2})$ (cf., expression for v). Therefore,

4.7. <u>Assertion</u>. $\varphi_{12}(n_{-1}tn_{-2}) \in \mathfrak{a}(\Gamma)$.

Next we have

$$(n_1tn_1 + n_{-1}t^{-1}n_{-1})\ n_{-1}tn_{-2} = t^2n_{-1} + n_{-3}\begin{pmatrix} 0 & -2^{-1} \\ 2 & 0 \end{pmatrix} n_{-4}$$

Since this latter expression has only one common term with Γ^2, namely t^2n_{-1}, we have $\varphi_{12}(t^2n_{-1})$, and by 4.4 also $\varphi_{12}(n_{-1})$ belongs to $\mathfrak{a}(\Gamma)$.

4.8. <u>Assertion</u>. $\varphi_{12}(n_{-1}) \in \mathfrak{a}(\Gamma)$.

From 4.7 and 4.8 it follows that $\varphi_{12}(t) \in \mathfrak{a}(\Gamma)$. Since t and n_{-1} generate $SL(2,\mathbf{F}_p)$, we have

4.9. <u>Assertion</u>. $\varphi_{12}(SL(2,\mathbf{F}_p)) \subset \mathfrak{a}(\Gamma)$.

Now note that $\Gamma \cdot \varphi_{12}(t)$ and $\varphi_{12}(n_1)[\Gamma \cdot \varphi_{12}(t)]\ \varphi_{12}(n_{-1})$ have only the following part in common:

$$\varphi_{12}(1 + n_c + n_1).$$

Since $1, \varphi_{12}(n_1) \in \mathfrak{a}(\Gamma)$, we have $\varphi_{12}(n_c) \in \mathfrak{a}(\Gamma)$. By Dixon's theorem $(SL(2,\mathbf{F}_q)$ is generated by $\begin{pmatrix} 1 & 0 \\ 1 & 1 \end{pmatrix}$ and $n_c)$ and we have

4.10. <u>Proposition</u>. $\varphi_{12}(SL(2,\mathbf{F}_q)) \subset \mathfrak{a}(\Gamma)$.

Therefore

$$\bar{w} \in \mathfrak{a}(\Gamma)$$

Note that w is the only term common to $\varphi_{12}(n_{-1})\ \bar{w}$ and $\varphi_{12}(n_{-4})\ \bar{w}$, whence

4.11. <u>Assertion</u>. $w \in \mathfrak{a}(\Gamma)$.

To get our theorem, note that $w^i \varphi_{12}(SL_2(\mathbf{F}_q))\, w^{-i} = \varphi_{12+i}(SL_2(\mathbf{F}_q))$ for $i = 0, 1, 2, \ldots, n-1$, and that these groups $\varphi_{1j}(SL_2(\mathbf{F}_q))$, $j = 2, \ldots, n+1$, generate $SL_{n+1}(\mathbf{F}_q)$.

AC. ONE MORE EXAMPLE OF A CELL WITH ONE GENERATOR.

The two preceding sections were, in fact, dedicated to a proof that in some cases a cellular algebra (it was $\mathbf{Z}[G]$) has one generator (as a cellular algebra) and that one can take a simple graph as such a generator. In this section, we consider one more example of this kind.

1. Again let $\mathbf{F}_q$ be the finite field with q elements, G an absolutely almost simple connected and simply connected algebraic group, defined and split over $\mathbf{F}_q$ (cf., e.g., [Bo 1]). Let T be a maximal torus of G, defined and split over $\mathbf{F}_q$, B a Borel subgroup of G containing T. Further let $G(\mathbf{F}_q)$ be the set of the $\mathbf{F}_q$-rational points of G, $N_{G(\mathbf{F}_q)}(T(\mathbf{F}_q))$ the normalizer of $T(\mathbf{F}_q)$ in $G(\mathbf{F}_q)$, $W = N_{G(\mathbf{F}_q)}T(\mathbf{F}_q))/T(\mathbf{F}_q)$ the Weyl group of G with respect to T.

As an example, one can take $G = SL_{n+1}$. Then $G(\mathbf{F}_q) = SL_{n+1}(\mathbf{F}_q)$ is the group of the preceding section, T is the set of diagonal matrices in G, $N_{G(\mathbf{F}_q)}(T(\mathbf{F}_q))$ is the set of monomial matrices with entries in $\mathbf{F}_q$. Then W is isomorphic to the symmetric group $\mathrm{Sym}(n + 1)$.

We want to show that the centralizer ring

$$\mathfrak{A} = \mathfrak{Z}(G(\mathbf{F}_q), G(\mathbf{F}_q)/B(\mathbf{F}_q))$$

has one generator Γ which is a simple graph: $\mathfrak{A} = \mathfrak{A}(\Gamma)$.

<u>Remark</u>. If it were known that $\mathrm{Aut}\,\mathfrak{A} = G(\mathbf{F}_q)$, we would have the stronger assertion, that there exists a simple graph Γ such that $\mathrm{Aut}\,\Gamma \simeq G(\mathbf{F}_q)$, $|\Gamma| = |G(\mathbf{F}_q)/B(\mathbf{F}_q)|$, $\mathfrak{A}(\Gamma) = \mathfrak{A}$.

2. It is known that $\mathfrak{A}$ is isomorphic as an algebra to $\mathbf{Z}[W]$, where W is the Weyl group of G (see [Bo 1], [Iw 1], [Yo 1]).

It is also known that double cosets $S_w = B(\mathbf{F}_q)\, wB(\mathbf{F}_q)$, $w \in W$, form a standard basis of $\mathfrak{A}$ and for fundamental reflections $w_1, \ldots, w_\ell$, one has the following relations (see [Iw 1], [Yo 1]):

$$S_{w_i} S_w = S_{w_i w} \quad \text{if} \quad \ell(w_i w) > \ell(w)$$

$$S_w S_{w_i} = S_{ww_i} \quad \text{if} \quad \ell(ww_i) > \ell(w)$$

$$S_w S_{w_i} = qS_{w_i w} + (q - 1)\, S_w \quad \text{if} \quad \ell(ww_i) < \ell(w)$$

$$S_{w_i} S_w = qS_{ww_i} + (q - 1)\, S_w \quad \text{if} \quad \ell(w_i w) < \ell(w)$$

Here $\ell(w)$ is the length of a shortest expression of w through w_i.

Remark. The following considerations are easily extendable to the case of quasi-split groups and Ree groups of type F_4. Such generalization influences only the structure constants in the expressions of $S_w S_{w_i}$ in the basic elements.

3. Theorem. Let $\mathfrak{A} = \mathcal{J}(G(\mathbf{F}_q), (G/B)(\mathbf{F}_q))$, $q > 2$ and $\operatorname{rg} G \geq 2$. Then there exists a simple graph

$$\Gamma \in \mathfrak{A}, \text{ such that } \mathfrak{A}(\Gamma) = \mathfrak{A}$$

Proof. Let Δ be a system of simple roots of G. To begin with, let us consider the case $|\Delta| \geq 3$. Let us choose two roots $\alpha_r, \alpha_m \in \Delta$ such that α_r and α_m generate a subsystem of type A_2 in the root system of G. Let $\Delta = \{\alpha_i\}$ and suppose that w_i is the reflection in α_i. Put $S_i = S_{w_i}$ and

$\Gamma = S_r + S_{w_r w_m} + S_{w_m w_r} + \Sigma_{i \neq r,m} S_i$. One has

$S_i^2 = q \cdot 1 + (q - 1)\, S_i$

$S_i S_r = S_{w_i w_r}, S_r S_i = S_{w_r w_i},$

$S_i S_{w_r w_m} = S_{w_i w_r w_m}$ if $i \neq r,m,$

$S_i S_{w_m w_r} = S_{w_i w_r w_m}$ if $i \neq r,m,$

$S_{w_m w_r} S_i = S_{w_m w_r w_i}$ if $i \neq r,m,$

$S_{w_r w_m} S_i = S_{w_r w_m w_i}$ if $i \neq r,m$,

$S_r S_{w_r w_m} = qS_m + (q-1)\ S_{w_r w_m}$

$S_r S_{w_m w_r} = S_{w_r w_m w_r}$

$S_{w_r w_m} S_r = S_{w_r w_m w_r}$

$S_{w_m w_r} S_r = qS_m + (q-1)\ S_{w_m w_r}$

$(S_{w_m w_r})^2 = (S_m S_r S_m)\ S_r = S_r S_m S_r S_r = qS_r S_m + (q-1)\ S_r S_m S_r$

(here we used the relation $S_r S_m S_r = S_m S_r S_m$ which follows from the corresponding relation in W).

$(S_{w_r w_m})^2 = (S_r S_m S_r)\ S_m = (S_m S_r S_m)\ S_m$

$= qS_m S_r + (q-1)\ S_m S_r S_m =$

$= qS_m S_r + (q-1)\ S_r S_m S_r$

$S_{w_r w_m} S_{w_m w_r} = S_r S_m S_m S_r$

$= qS_r^2 + (q-1)\ S_r S_m S_r = q^2 \cdot 1 + q(q-1)\ S_r + (q-1)\ S_r S_m S_r$

$S_{w_m w_r} S_{w_r w_m} = q^2 \cdot 1 + q(q-1)\ S_m + (q-1)\ S_r S_m S_r$

Therefore we have

$\Gamma^2 = (q + 2q^2) \cdot 1 + (q^2 - 1)\ S_r + (q^2 + q)\ S_m + (2q - 1)\ S_m S_r$

$+ (4q - 2)\ S_r S_m S_r + (q - 1)\ \Sigma_{i \neq r,m} S_i + (\ldots).$

All summands in the parentheses have coefficient 1 or 2 (namely, if $w_i w_j = w_j w_i$, then $S_{w_i w_j}$, $i, j \neq m, r$, has coefficient 2).

Since $q > 2$, we see that $q^2 - 1$, $q^2 + q$, $2q - 1$, $4q - 2$, $q - 1$ are all distinct. Also $q^2 - 1$, $q^2 + q$, $2q - 1$, $4q - 2$ are greater than 2 (if $q > 2$). Therefore, $S_r, S_m, S_{m_r}, S_r S_m S_r \in \mathfrak{A}(\Gamma)$. But then $S_{r_m} \in \mathfrak{A}(\Gamma)$, whence $\Sigma_{i \neq r,m} S_i \in \mathfrak{A}(\Gamma)$. Next we can apply the same reasoning as in the case of $Z[\mathrm{Sym}(n)]$. Namely, we can take one of the S_r or S_m such that (denote our choice by S):

$$S(\Sigma_{i \neq r,m} S_i)\, S = S\, S_j\, S + \Sigma_{i \neq r,m,j} S_i$$

whence $S_j \in \mathfrak{A}(\Gamma)$ and so forth.

Let us now consider the case $|\Delta| = 2$, $\Delta = \{\alpha_1, \alpha_2\}$. If G has type A_2, then all preceding computations without any alteration or new comment lead us to our assertion. When G is not of type A_2 (that is, it is of type B_2 or G_2), we put

$$\Gamma = S_1 + S_1 S_2 + S_2 S_1$$

Since G is not of type A_2, one has

$$(S_1 S_2)(S_1 S_2) = S_{w_1 w_2 w_1 w_2}$$

$$(S_2 S_1)(S_2 S_1) = S_{w_2 w_1 w_2 w_1}$$

Therefore

$$\begin{aligned} \Gamma^2 &= q \cdot 1 + (q - 1)\, S_1 + qS_2 + (q - 1)\, S_1 S_2 \\ &+ S_1 S_2 S_1 + S_1 S_2 S_1 + qS_2 + (q - 1)\, S_2 S_1 \\ &+ S_1 S_2 S_1 S_2 + S_2 S_1 S_2 S_1 + q^2 1 + q(q - 1)\, S_1 \\ &+ (q - 1)\, S_1 S_2 S_1 + q^2 1 + q(q - 1)\, S_2 + (q - 1)\, S_2 S_1 S_2 \end{aligned}$$

This again implies that $S_1, S_2 \in \mathfrak{A}$ (since these summands have the greatest and unequal coefficients). Thus in the case $|\Delta| = 2$, our assertion also follows.

AD. DEEP CONSTANTS

1. We consider here formations which arise in one of the possible definitions of depth compare, O 6.3. The discussions of this section are rather fragmentary and their aim is to point out relations which one obtains with the strengthening of a definition of a stationary graph. Analogous relations are considered in [He 1].

We examine here only the case of a cell; the extension to general stationary graphs can be obtained without difficulty but leads to a much more complex exposition, compare Section D with E 4.

2. Let $X = (x_{ij})$ be a stationary graph, $X = \Sigma_{i\in I} x_i e_i$. We say that the edge x_{ab} is of type k if $x_{ab} = x_k$.

Let $(V(X))^m$ be the set of all ordered m-tuples of vertices. Let $(a, b, c_1, \ldots, c_m) \in (V(X))^{m+2}$. We say that this set is of type $(k; i_1, \ldots, i_m; j_1, \ldots, j_m; A)$, where $k, i_p, j_p \in I$, $A = (s_{pq})$ is $(m\times m)$-matrix, $s_{pq} \in I$, $s_{pq} = s'_{qp}$, if $x_{ab} = x_k$, $x_{a,c_p} = x_{i_p}$, $x_{c_p,c_q} = x_{s_{pq}}$. We shall write $\vec{i}$ in place of $(i_1, \ldots, i_m)$ and $\vec{j}$ in place of $(j_1, \ldots, j_m)$. Let us note that it would be more convenient to consider $\vec{i}$ as a row-vector and $\vec{j}$ as a column vector. In this section we use the following notation: If $\vec{i} = (i_1, \ldots, i_m)$, then $\vec{i}' = (i'_1, \ldots, i'_m)$, $\vec{i}^t$ is the column-vector with the same coordinates. If A is a matrix, then A^t is its transpose.

3. Let us consider the class of stationary graphs such that for any edge (a,b) of type k and for any $m \in [1,n]$ the number of all sets $(a,b; c_1, \ldots, c_m)$ of any given type $(k; \vec{i}; \vec{j}; A)$ is the same (and does not depend on the choice of the edge (a,b) of type k). (Note that these are ordered sets, possibly with repetitions.) We denote this number by $a^k_{\vec{i},\vec{j},A}$.

If G is a permutation group on $V(X)$ and $X = X(\mathcal{J}(G,V(X)))$, then G acts transitively on edges of a given type (cf., F 4.1). Therefore, such graphs X satisfy the above condition.

4. Let us state some relations for the numbers $a^k_{\vec{i},\vec{j},A}$ in the case of a cell. Let

$e_0 = E_n$.

Put $A = (s_{pq})$. We shall sometimes write A in the form

$$A = \begin{pmatrix} & B & & \vec{s}_1 \\ & & & \cdot \\ & & & \cdot \\ & & & \cdot \\ & & & \vec{s}_r \\ \vec{s}_1'^{\,t}, & \ldots, & \vec{s}_r'^{\,t} & C \end{pmatrix}$$

where B is $(r \times r)$-matrix, C is $(m - r \times m - r)$-matrix, s_i is a vector of length $m - r$, $r < m$.

Furthermore, we write:

$\hat{A}_r$ for the matrix obtained from A by deleting the r-th row and r-th column,

$(i_1, \ldots, \hat{i}_r, \ldots, i_m)$ for the vector with deleted r-th component,

$\vec{i}_1, \vec{i}_2$ for the vector whose components are components of $\vec{i}_1$ followed by those of $\vec{i}_2$.

If $g \in \mathrm{Sym}(m)$, then

$$g\vec{i} = (i_{g(1)}, i_{g(2)}, \ldots, i_{g(m)})$$

<u>Theorem</u>. The following relations hold.

4.1. $a^k_{\vec{i},\vec{j},A} = a^{k'}_{\vec{j}',\vec{i}',A}$.

4.2. $a^k_{g\vec{i},g\vec{j},g^{-1}Ag} = a^k_{\vec{i},\vec{j},A}$, $g \in \mathrm{Sym}(m)$.

4.3. $a^0_{\vec{i},\vec{j},A}$

$= (\prod_{k=1}^m \delta_{i_k j_k'}) \, n_{i_r} a^{i_r}_{i_1, \ldots, \hat{i}_r, \ldots, i_m; s_{r1}, \ldots, \hat{s}_{rr}, \ldots, s_{rm}; \hat{A}_r}$ for all $r \in [1,m]$.

4.4. $\Sigma_{\vec{i}} a^k_{\vec{i},\vec{j},A} = a^0_{\vec{j},\vec{j}',A}$.

4.5. $\sum_{\vec{j}_2,\vec{s}_2,\ldots,\vec{s}_r} a^k_{\vec{i}_1\vec{i}_2;\vec{j}_1\vec{j}_2;A} = a^k_{\vec{i}_1,\vec{j}_1,B}\, a^{i_1}_{\vec{i}_2,\vec{s}'_1,C}$.

4.6. $\sum_{\vec{i}_2,\vec{j}_2,\vec{s}_3,\ldots,\vec{s}_r} a^k_{\vec{i}_1\vec{i}_2;\vec{j}_1\vec{j}_2;A} = a^k_{\vec{i}_1,\vec{j}_1,B}\, a^{s_{12}}_{\vec{s}_1,\vec{s}'_2,C}$.

4.7. $\sum_{\vec{s}_1,\vec{s}_2,\ldots,\vec{s}_r} a^k_{\vec{i},\vec{j},A} = a^k_{\vec{i}_1,\vec{j}_1,B} \cdot a^k_{\vec{i}_2,\vec{j}_2,C}$.

<u>Proofs</u>. They are elementary and analogous to the geometrical proofs of properties of the numbers a^k_{ij}, compare D 4 c 1 - c 8.

Let (a,b) be an edge of type k and $(a,b;\ c_1,\ \ldots,\ c_m)$ a set of type $(k;\ \vec{i},\vec{j},A)$.

Then the set $(b,a;\ c_1,\ \ldots,\ c_m)$ is of type $(k';\ \vec{i}',\vec{j}',A)$. This proves 4.1.

If $g \in \mathrm{Sym}(m)$ then $(a,b;\ c_{g(1)},\ \ldots,\ C_{g(m)})$ is a set of type $(k;\ g\vec{i},\ g\vec{j},\ g^{-1}Ag)$ whence 4.2.

To prove 4.3, consider the case $a = b$. Then (since $a = b$) the number $a^0_{\vec{i},\vec{j},A}$ is zero unless $i_k = j'_k$ for all k. Suppose that $i_k = j'_k$ for all k. Then $(a,c_r;\ c_1,\ \ldots,\ \hat{c}_r,\ \ldots,\ c_m)$ is a set of type $(i_r;\ i_1,\ \ldots,\ \hat{i}_r,\ \ldots,\ i_m;\ s_{r1},\ \ldots,\ \hat{s}_{rr},\ \ldots,\ s_{rm},\ \hat{A}_r)$. On the other hand, let $d_1,\ \ldots,\ d_{n_{i_r}}$ be vertices of X such that (a,d_i) is of type i_r. Then by the assumption stated in the beginning of 3, there exists the same number

$$a^{i_r}_{i_1,\ldots,\hat{i}_r,\ldots,i_m;s_{r1},\ldots,\hat{s}_{rr},\ldots,s_{rm},\hat{A}_r}$$

of sets of type $(i_r;\ i_1,\ \ldots,\ \hat{i}_r,\ \ldots,\ i_m;\ s_{r1},\ \ldots,\ \hat{s}_{rr},\ \ldots,\ s_{rm},\ \hat{A}_r)$ on each of the n_{i_r} edges (a,d_i). Each such set can be considered as a set of type $(0;\ \vec{i},\vec{i}',A)$ on (a,a) and each set of type $(0;\vec{i},\vec{i}',A)$ gives rise to a set of new type in the manner described above. Therefore, 4.3 is proved.

The summation in 4.4 means that we have to consider all sets $(a,b;\ c_1,\ \ldots,\ c_m)$ such that the types of edges (a,c_r) are arbitrary. It is the same as to consider all sets $(a,b;c_1,\ldots,c_m)$ for which only the types of the edges

(b,c_r) and (c_s,c_t) are fixed. This is the same as to consider the sets $(b,b;\ c_1,\ \dots,\ c_m)$. Since the number of these sets is $a^0_{\vec{j},\vec{j}',A}$, 4.4 is proved.

The summation in 4.5 means that we have to consider all sets $(a,b;\ c_1,\ \dots,\ c_m)$ such that the types of edges $(b,c_{r+1}),\ \dots,\ (b,c_m);\ (c_s,c_t),\ s \in [2,r],\ t \geq r+1$, are arbitrary; $(a,c_s),\ s \in [1,r]$, are of type i_s; $(b,c_s),\ s \in [1,r]$, are of type j_s, $(c_q,c_t),\ q \notin [2,r]$ or $t \leq r$, are of type $s_{q,t}$. This means that we have to consider all configurations (there are $a^k_{\vec{i}_1,\vec{j}_1,B}$ of them) $(a,b;\ c_1,\ \dots,\ c_r)$ of type $(k;\ \vec{i}_1,\vec{j}_1,B)$ and on the edge (a,c_1) of each of them all configurations of type $(i_1;\ \vec{i}_2,\vec{s}\,'_1,C)$ (there are $a^{i_1}_{\vec{i}_2,\vec{s}_1,C}$ of them). So the entire number is the product of these numbers, which proves 4.5.

The summation in 4.6 means that we have to consider all sets $(a,b;\ c_1,\ \dots,\ c_m)$ such that only types of $(a,c_i),\ i \leq r;\ (b,c_i),\ i < r;\ (c_s,c_t),\ s \notin [3,r]$ or $t \leq r$, are fixed. This means that we have to consider all configurations of type $(k;\ \vec{i}_1,\vec{j}_1,B)$ (there are $a^k_{\vec{i}_1,\vec{j}_1,B}$ of them) $(a,b;\ c_1,\dots,\ c_r)$ on the edge (a,b) and on the edge (c_1,c_2) of them all configurations of type $(s_{12},\vec{s}_1,\vec{s}\,'_2,C)$ (there are $a^{s_{12}}_{\vec{s}_1,\vec{s}_2,C}$ of them). So the entire number is the product of these numbers, which proves 4.6.

The summation in 4.7 means that we have to consider all sets $(a,b;\ c_1,\dots,\ c_m)$ such that only types of $(a,c_i),\ i \in [1,m];\ (b,c_i),\ i \in [1,m];\ (c_q,c_t),\ q \leq r$, $t \geq r+1$, are fixed. This means that we have to consider all configurations of type $(k;\ \vec{i}_1,\vec{j}_1,B)$ on the edge (a,b) (there are $a^k_{\vec{i}_1,\vec{j}_1,B}$ of them) and all configurations of type $(k;\ i_2,j_2,C)$ on the edge (a,b) (there are $a^k_{\vec{i}_2,\vec{j}_2,C}$ of them). So the entire number is the product of these numbers, which proves 4.7.

5. If $m = 1$, the constants of this section are the structure constants a^k_{ij} of the algebra $\mathfrak{a}(X)$. It is interesting to note that all relations D 4 c 1 - c 6 on a^k_{ij} are corollaries of the relations 4.1 - 4.6 for $m = 1$ and/or $m = 2$.

E.g., $n_i a^i_{jk} = n_j a^{j'}_{ki'}$ follows from 4.3 and 4.1:

$$a^{0}_{ij;i'j'} \begin{pmatrix} 0 & k \\ k' & 0 \end{pmatrix} = n_i a^{i}_{jk} = n_j a^{j}_{ik'} = n_j a^{j'}_{ki'}$$

One more example:

$$\Sigma_s\, a^{s}_{ij} a^{k}_{s\ell} = \Sigma_s\, a^{k}_{is} a^{s}_{j\ell}$$

follows from 4.5 and 4.2:

$$\Sigma_{t,s}\, a^{k}_{it,s\ell} \begin{pmatrix} 0 & j \\ j' & 0 \end{pmatrix} = \Sigma_s\, a^{k}_{is} a^{s}_{j\ell} = \Sigma_t\, a^{k}_{ij} a^{k}_{t\ell}$$

6. (Compare D4 c 9, c 10). Let

$$B = \begin{pmatrix} 0 & \vec{j} \\ \vec{j}' & A \end{pmatrix} .$$

6.1. Lemma. $a^{0}_{k,\vec{i};k',\vec{i}',B} = n_k a^{k}_{\vec{i},\vec{j},A}$

Proof follows directly from 4.3.

6.2. Corollary. $n_k a^{k}_{\vec{i},\vec{j},A} = n_{i_r} a^{i_r}_{k,i_1,\ \ldots,\ \hat{i}_r,\ \ldots,\ i_m;\ j'_r, s_{r,1},\ \ldots,\ \hat{s}_{rr},\ \ldots,\ s_{rm};\hat{B}_r}$

Proof follows from 6.1 and 4.3.

6.3. Theorem. Let $N_{\vec{i},\vec{j},A}$ be the least common multiple of all numbers $n_{i_p}, n_{j_p}, n_{s_{pq}}$. Then $N_{\vec{i},\vec{j},A}$ divides $n_k a^{k}_{\vec{i},\vec{j},A}$.

Proof. By 6.2, $n_k a^{k}_{\vec{i},\vec{j},A}$ is a multiple of all n_{i_r}. By 6.2 and 6.1 it is also a multiple of all n_{j_r}. Applying the same reasoning to the right side of 6.2, we see that our number is a multiple of all $n_{s_{pq}}$, as required.

6.4. Remark. It follows from 6.3 that $a^{k}_{\vec{i},\vec{j},A}$ is a multiple of all numbers

$n_k \cdot (n_k, N_{\vec{i},\vec{j},A})^{-1}$. Applying this consideration to the right side of 6.2, one can strengthen Theorem 6.3. Specifically, one can define $N_{\vec{i},\vec{j},A}$ recursively as the least common multiple of all numbers

$n_{i_r}^2 (n_{i_r}, N_{k,i_1,\ldots,\hat{i}_r,\ldots,i_m;j_r',s_{r1},\ldots,\hat{s}_{rr},\ldots,s_{rm};\hat{B}_r})^{-1}$. This strengthening of 6.3 also is not final since one can iterate this consideration.

AE. ALGEBRAIC INVARIANTS OF FINITE GRAPHS

1. Many mathematicians, myself included, who have an algebraic background, say, when told about the problem of graph isomorphism: "What is the question? Invariant polynomials surely distinguish graphs up to isomorphism."

1.1. They allude to the following result from the algebraic geometry (cf., [Se 4]):

Let M be an affine algebraic manifold over a field k, $k[M]$ be the ring of regular functions on M. Suppose that a finite group G acts on M (and preserves the algebraic structure). Let $R = k[M]^G$ be the ring of invariants under G regular functions. Then R distinguishes orbits of G on M and M/G is the affine manifold whose ring of regular functions is R.

This means that two points of M lie on the same orbit of G if and only if the values of any function from R are the same at those two points.

1.2. In our case, M is the space of all matrices of order n (or of all symmetric matrices, or of all symmetric matrices with zero diagonal) and $G = \mathrm{Sym}\ n$ acts on M in the following manner:

$$A \longrightarrow g\,A\,g^{-1}$$

According to the Hilbert's Theorem, the ring of invariants has a finite generator set. Hence the approach stated above is not infinite.

1.3. However, if $f_1, \ldots, f_N$ are the generators of the ring of invariants, one should compute all functions f_i to establish the isomorphism or non-isomorphism. But f_i can be rather complicated and the number N can be large.

1.4. The aim of the present section is to exhibit and to interpret some of the complications which were encountered during an attempt to use the invariants. It seems that these complications are of the same nature as those of Sections M, AD.

2. We shall show below how the values of the invariants of simple graphs (that is, of its adjacency matrix) are interpreted in geometrical terms.

2.1. To every monomial m of the form

$$m = \prod a_{ij}$$

in the matrix entries we associate a graph $\Gamma(m)$ in the following manner:

$\Gamma(m)$ has the edge (i,j) if and only if a_{ij} enters in m.

Then the number of edges of $\Gamma(m)$ is the degree of m; the number of vertices of $\Gamma(m)$ is the number of distinct indices of a_{ij} entering in m.

If Γ is a simple graph, then the value of m on $A(\Gamma)$ is 1 if $\Gamma(m)$ (with the given numeration of vertices) is a subgraph of Γ, and is 0 if $\Gamma(m) \not\subset \Gamma$.

2.2. Now let f be an invariant polynomial for the group Sym n on the space of the symmetric $n \times n$-matrices with zero diagonal. It is easy to see that f is a linear combination of invariants of the form

$$f(m) = \Sigma\, m^g$$

where g runs over a set of coset representatives of Sym n by a subgroup, fixing m. Therefore, it can be assumed that f has the above form, i.e., $f = f(m)$.

Then the value of f on $A(\Gamma)$ is evidently equal to the <u>number of embeddings</u> of (the abstract graph) $\Gamma(m)$ in Γ considered up to isomorphism.

2.3. <u>Remark</u>. It can be seen from the above argument that the Theorem 1.1 is evident in our case, since to the whole graph Γ there corresponds some monomial of degree equal to the number of edges of Γ, and to this monomial there corresponds the invariant which itself completely determines the isomorphism class of Γ.

2.4. By 2.3 to distinguish graphs with n vertices, it is sufficient to consider only invariants of degrees $\leq n^2$. However, all those invariants should not be considered. It is sufficient to construct a basis of invariants. But this basis contains perhaps too many invariants and many of them have (also perhaps) a high degree (e.g., degree n). Computation of an invariant of degree t requires, a priori, $t \cdot \binom{n^2}{t}$ actions. This shows that the possibility to use algebraic invariants requires further research.

2.5. Remark. Note that the values of some invariants of high degree on $A(\Gamma)$ can be determined from values of simpler ones without use of algebraic relations. Define for every monomial m another monomial $\bar{m}$ in the following manner:

$\bar{m}$ is the product of the matrix entries entering in m and taken in the first power, that is, if $m(A) = \prod a_{ij}^{m_{ij}}$, then $\bar{m}(A) = \prod a_{ij}^{t(m_{ij})}$ where

$t(m_{ij}) = 1$ if $m_{ij} > 0$ and $t(m_{ij}) = 0$ if $m_{ij} = 0$.

If $A = A(\Gamma)$, then evidently $\bar{m}(A) = m(A)$. If $f = f(m)$, $\bar{f} = f(\bar{m})$, then $f(A) = d \cdot \bar{f}(A)$, where d is the index of the group fixing m in the group fixing $\bar{m}$.

This shows that in order to study isomorphisms of simple graphs (or more generally of graphs without multiple edges) it is sufficient to consider the invariants $\bar{f}$.

3. Remarks of the preceding subsection are easily extended to the general case. Indeed, suppose that we have to establish whether two $n \times n$-matrices A and B belong to the same orbit of $\mathrm{Sym}\, n$. Replace the pair $\{A,B\}$ by $X(\{A,B\}) = \{X,Y\}$ (simultaneous stabilization, cf. M 4).

Let us now indicate some analogues of the considerations of the preceding subsection.

3.1. To every monomial in the variables entering in X and Y, let us associate a graph (in the sense of Section C) in the same manner as in 2.1. Then the monomial m in the matrix entries determines an equivalence class of graphs (cf., C 2). The value of $f(m)$ on X determines the number of embeddings (up to isomorphism) of every representative (up to isomorphism) of that equivalency class into X (an analogue of 2.2).

3.2. As an analogue of 2.5, the following schema is proposed.

In place of the invariants of degree r we shall consider elements of the tensor product of d distinct copies of the space of matrices (recall that invariants of degree r are elements of the r-th symmetric power of the dual space of the space of matrices). It frees us from the necessity to substitute m by $\bar{m}$ but leads to difficulties, one of which is the fact that our new object is not a ring.

4. Let us finally give a few <u>examples</u>.

Let $A = (a_{ij})$ be a matrix.

4.1. A basis of invariants of degree 1.

a) $\Sigma_{i\neq j}\, a_{ij}$

b) $\Sigma_i\, a_{ii}$

4.2. A basis of invariants of degree 2.

a) $\Sigma_{i>j}\, a_{ii}\, a_{jj}$

b) $\Sigma_i\, a_{ii}^2$

c) $\Sigma_{i\neq j}\, a_{ii}\, a_{ij}$

d) $\Sigma_{i\neq j, j\neq k, i\neq k}\, a_{ii}\, a_{jk}$

e) $\Sigma_{i\neq j}\, a_{ij}^2$

f) $\Sigma_{i\neq j}\, a_{ij}\, a_{ji}$

g) $\Sigma_{i\neq j, j\neq k, i\neq k}\, a_{ij}\, a_{jk}$

h) $\Sigma_{i\neq j, j\neq k, i\neq k}\, a_{ij}\, a_{ik}$

i) $\Sigma_{i\neq j, j\neq k, k\neq l, i\neq k, i\neq l, j\neq l}\, a_{ij}\, a_{kl}$

AG. CONJECTURES.

Below we state some conjectures and indicate directions of research which are now of interest for us.

1. <u>Conjecture</u>. (Arlazarov). Let Γ be a graph, $n = |\Gamma|$. The algorithm of Section R finishes its work in $n^{C \log n}$ steps.

It is interesting to note that usually when an algorithm is given, the estimates of its speed are obtained relatively easily. In the case under consideration for many examples the algorithm finishes its work "momentarily;" however, no good estimate is obtained. Hypothesis 1 is close and, possibly, equivalent to the assumption:

If $X = (X_{ij})$ is a stationary graph of depth $(\log |X|)$ in some sense (cf., O 6), then the sets $V(X_{ii})$ are orbits of Aut X.

One of the obstructions to the proof of the estimate is the necessity to pay special attention to correct graphs.

2. <u>Question</u>. Construct a lower bound on the number of steps required to establish isomorphism or non-isomorphism of graphs.

Perhaps, in order to find such bounds one should be able to compute the number of stationary graphs of the given depth k which have the same "structure constants" for every depth $\leq k$.

It is interesting to note that all algorithms for establishing isomorphism known to us are algorithms of canonization; that is, they reorder vertices of the first graph and of the second graph, and then compare results. The use of algebraic invariants (cf., Section AE) just solves the isomorphism problem (without canonizing). Difficulties arising in that direction were indicated in Section AE.

3. <u>Question</u>. Find an algorithm of construction of cells (or of strongly regular graphs) which will construct exactly one representative of each isomorphism class.

Such an algorithm, if it exists, has to use deep structural properties of cells. Seidel equivalence, descent-ascent (cf., Section V) are examples of the existence of large common parts of strongly regular graphs. Some other cases of closeness are pointed out in Section V.

4. Question. Does the center (as algebra) of a cell form a cell?

If it were true, more information on a_{ij}^{k} and on the existence would be obtained. Besides, such an assertion would have an independent interest. For some centralizer rings it was proved in [Ta 3]. Possibly by the method of [Ta 3] at least a proper sub-cell of a cell can be constructed. If such a process would stop, the cells where it does not give proper subalgebra are of special interest.

5. Question. Extend the results of R. E. Block [Bl 1], [Bl 2] to cellular algebras.

Those results have formulations which also make sense for cellular algebras. Their proof would give ample information about the structure of cellular algebras.

6. Question. Extend the Chowla-Bruck-Ryser theorem [Ha 3] to cellular algebras. A Hermitian form may be associated to every basic element (it is quadratic if that element is symmetric). Therefore, results of Hasse-Minkowski are applicable, in principle. The question is how to express the invariants of our forms in terms of the structure constants. In particular, the question arises whether they are expressible or not. Note, however, that if they are not expressible, a new invariant of a cellular algebra would be obtained.

BIBLIOGRAPHY

[Ad 1] Adelson-Velsky, G.M., Arlazarov, V.L., Bitman, A.R., Zhivotovsky, A.A., Uskov, A.V., On programming chess for computers (in Russian), Uspehi Math. Nauk 25 (1970), 222-260.

[Ad 2] Adelson-Velsky, G.M., Weisfeiler, B., Lehman, A.A., Faragev, I.A., On an example of a graph having no transitive automorphism group (in Russian), Doklady AN SSSR, 185 (1969).

[Ah 1] Ahrens, R.W., Szekeres, G., On a combinatorial generalization of 27 lines associated with a cubic surface, J. Austr. Math. Soc., 10 (1969), 485-492.

[Ai 1] Aigner, M., On characterization problem in graph theory, J. Comb. Theory, 6 (1969), 45-55.

[Al 1] Albert, A.A., Structure of algebras, Amer. Math. Soc. Coll. Publ., v. 24, N.Y., 1939.

[Al 2] Aliev, J.S.O., Seiden, E., Steiner triple systems and strongly regular graphs, J. Comb. Theory, 6 (1969), 33-39.

[Ar 1] Arlazarov, V.L., Zuev, I.I., Uskov, A.V., Faragev, I.A., (in Russian), Zhurnal Vychisl. Math. y Math. Physicy, 14 (1974), 737-743.

[Ar 2] Arlazarov, V.L., Lehman, A.A., Rosenfeld, M.Z., Computer-aided construction and analysis of graphs with 25, 26, and 29 vertices (in Russian), Preprint, Institute of Control Problems, Moscow, 1975, (58 pages).

[As 1] Aschbacher, M., The non-existence of rank 3 permutation groups of degree 3250 and subdegree 57. J. algebra, 19 (1971), 538-540.

[Ba 1] Bauman, H., Computer program for LINCO System, J. Chem. Docum., 5 (1965), 14-23.

[Ba 2] Bauersfeld, G., Essmann, Ch., Löhle, H., Algorithmus zur Feststellung der Isomorphie von endlicher, zusammenhängenden Graphen, Computing 11 (1973), 159-168.

[Be 1] Benson, C.T., Losey, N.E., On a graph of Hoffman and Singleton, J. Comb. Theory, 11B (1971).

[Be 2] Berztiss, A.T., A backtrack procedure for isomorphism of directed graphs, J. of ACM, 20 (1973), 365-377.

[Bl 1] Block, R.E., On the orbits of collineation groups, Math. Z., 96 (1967), 33-49.

[Bl 2] Block, R.E., On automorphism groups of block designs, J. Comb. Theory, 5 (1968), 293-301.

[Bo 1] Borel, A., Tits, J., Groupes reductifs, Publ. Math. IHES, no. 27, 1965, 55-152.

[Bo 2] Bose, R.C., Strongly regular graphs, partial geometries and partially balanced designs, Pacific J. Math., 13 (1963), 389-419.

[Bo 3] Bose, R.C., Clatworthy, W.H., Some classes of partially balanced designs, Ann. Math. Stat., 26 (1955), 212-232.

[Bo 4] Bose, R.C., Dowling, T.A., A generalization of Moore graphs of diameter 2, J. Comb. Theory, 11 (1971), 213-226.

[Bo 5] Bose, R.C., Laskar, R., A characterization of tetrahedral graphs, J. Comb. Theory, 3 (1967), 366-385.

[Bo 6] Bose, R.C., Mesner, D.M., On linear associative algebras corresponding to association schemes of partially balanced designs, Ann. Math. Sat., 30 (1959), 21-38.

[Bo 7] Bose, R.C., Shimamoto, T., Classification and analysis of partially balanced incomplete block designs with two associate classes, J. Amer. Stat. Assn., 47 (1952), 151-184.

[Bo 8] Bose, R.C., Shrikhande, S.S., Graphs in which each pair of vertices is adjacent to the same number of other vertices, Studia Sci. Math. Hung., 5 (1970), 181-196.

[Bo 9] Bose, R.C., Shrikhande, S.S., Some further constructions for $G(d)$ graphs, Studia Sci. Math. Hung., 6 (1971), 127-132.

[Br 1] Bruck, R.H., Finite nets II, Uniqueness and imbedding, Pacific J. Math., 13 (1963), 421-457.

[Br 2] Brudno, A.L., Branches and boundaries for search reduction, (in Russian), Problemy kybernetiki, 10 (1963), 141-150.

[Ca 1] Cameron, P.J., Permutation groups with multiply transitive suborbits, Proc. London Math. Soc., III-Ser., 25 (1972), 427-440.

[Ca 2] Cameron, P.J., Goethals, J.M., Seidel, J.J., Shult, E.E., Line graphs, root systems and elliptic geometry, Preprint, 1975, to appear J. of Algebra.

[Ch 1] Chang Li-chien, The uniqueness and nonuniqueness of the triangular association schemae, Sci. Record, New Ser., 3 (1959), 604-613.

[Ch 2] Chang Li-chien, Association schemes of partially balanced designs with parameters $v = 28$, $n_1 = 12$, $n_2 = 15$, and $p_{11}^2 = 4$, Sci. Record, New Ser., 4 (1960), 12-18.

[Ch 3] Chao Chang-Yun, On a theorem of Sabidussi, Proc. AMS., 15 (1964), 291-292.

[Ch 4] Chao Chang-Yun, On groups and graphs, Trans. AMS, 118 (1965), 488-497.

[Co 1] Connor, W.S., The uniqueness of the triangular association schemes, Ann. Math. Stat., 29 (1958), 262-266.

[Co 2] Connor, W.S., Clatworthy, W.H., Some theorems for partially balanced designs, Ann. Math. Stat., 25 (1954), 100-112.

[Co 3] Corneil, D.G., Gotlieb, C.C., An efficient algorithm for graph isomorphism, J. of ACM, 17 (1970), 51-64.

[Da 1] Davydov, E.G., On finite graphs and their automorphisms, (in Russian), Problemy kibernetiki, 17 (1966), 27-39.

[Da 2] Davydov, E.G., On automorphisms of unions of finite products of graphs, (in Russian), Kibernetika, 1968, no. 6.

[De 1] Delsarte, P., An algebraic approach to the association schemes of coding theory, Philips Res. Repts., Suppl. no. 10 (1973)

[De 2] Delsarte, P., The association schemes of coding theory, Combinatorics, Math. Centre Tract, 55, (Amsterdam 1974), 139-157.

[De 3] Dembowski, P., Finite geometries, Berlin-Heidelberg-N.Y., Springer, 1968.

[Dj 1] Djokovic, D.Z., Isomorphism problem for a special class of graphs, Acta. Math. Hung., 21 (1970), 267-270.

[Do 1] Doob, M., Graphs with a small number of distinct eigenvalues, Ann. N.Y. Acad. Sci., 175 (1970), no. 1, 104-110.

[Do 2] Doob, M., On characterizing certain graphs with four eigenvalues by their spectra, Linear algebra and appl., 3 (1970), 461-482.

[Fr 1] Frame, J.S., The double cosets of a finite group, Bull. AMS, 47 (1941), 458-467.

[Fr 2] Frame, J.S., Double cosets matrices and group characters, Bull. AMS, 49 (1943), 81-92.

[Fr 3] Frucht, R., Herstellung von Graphen mit vorgegebenen abstrakten Gruppe, Comp. Math., 6 (1938), 239-250.

[Ge 1] Gewirtz, A., Graphs with maximal even girth, Canadian J. Math., 21 (1969), 915-934.

[Go 1] Goethals, J.M., Seidel, J.J., Orthogonal matrices with zero diagonal, Canadian J. Math., 19 (1967), 1001-1010.

[Go 2] Goethals, J.M., Seidel, J.J., Strongly regular graphs derived from combinatorial designs, Canadian J. Math., 22 (1970), 449-471.

[Ha 1] Halin, R., Jung, H.A., Note on isomorphism of graphs, J. London Math. Soc., 42 (1967), 254-256.

[Ha 2] Hall, M., Jr., The theory of groups, N.Y., 1959.

[Ha 3] Hall, M., Jr., Combinatorial theory, Waltham-Toronto-London, 1967.

[He 1] Hemminger, R.L., On the group of a directed graph, Canadian J. Math., 18 (1966), 211-220.

[He 2] Hestenes, M.D., Higman, D.G., Rank 3 groups and strongly regular graphs, SIAM-AMS Proceedings, vol. 4, Providence, 1971, 141-160.

[Hi 1] Higman, D.G., Finite permutation groups of rank 3, Math. Z., 86 (1964), 145-156.

[Hi 2] Higman, D.G., Intersection matrices for finite permutation groups, J. algebra, 6 (1967), 22-42.

[Hi 3] Higman, D.G., Coherent configurations I, Rend. Sem. Padova, 44 (1970), 1-25.

[Hi 4] Higman, D.G., Characterization of families of rank 3 permutation groups by the subdegrees I, II, Arch. Math., 21 (1970), 151-156, 353-361.

[Hi 5] Higman, D.G., Schur relations for weighted adjacency algebras, Symp. Math. (Roma), vol. 13, London-N.Y., 1974, 467-477.

[Hi 6] Higman, D.G., Coherent configurations. Part I: Ordinary representation theory, Geom. Dedicata, 4 (1975), 1-32.

[Hi 7] Higman, D.G., Sims, C.C., A simple group of order 44, 353,000, Math. Z., 105 (1968), 110-113.

[Ho 1] Hoffman, A.J., On the exceptional case in a characterization of the arcs of a complete graph, IBM J. Res. Dev., 4 (1960), 487-496.

[Ho 2] Hoffman, A.J., On the uniqueness of the triangular association scheme, Ann. Math. Stat., 31 (1960), 492-497.

[Ho 3] Hoffman, A.J., On the polynomial of a graph, Amer. Math. Monthly, 70 (1963), 30-36.

[Ho 4] Hoffman, A.J., On the line graph of a complete bipartite graph, Ann. Math. Stat., 35 (1964), 883-885.

[Ho 5] Hoffman, A.J., On the line graph of a projective plane, Proc. AMS, 16 (1965), 292-302.

[Ho 6] Hoffman, A.J., Newman, M., Straus, E.G., Taussky, O., On the number of absolute points of a correlation, Pacific J. Math., 6 (1965), 83-96.

[Ho 7] Hoffman, A.J., Chaudhuri, D.K. Ray, On the line graph of a symmetric incomplete block design, Trans. AMS, 116 (1965), 238-252.

[Ho 8] Hoffman, A.J., Singleton, R.R., On Moore graphs with diameter 2 and 3, IBM J. Res. Dev., 4 (1960), 492-504.

[Ho 9] Hopcroft, J.E., Tarjan, R.E., A V log V algorithm for isomorphism of triconnected planar graphs, J. of Computer and Syst. Sci., 7 (1973), 321-323.

[Ho 10] Hopcroft, J.E., Wong, J.K., Linear time algorithm for isomorphism between planar graphs, 6th Annual ACM Symp. on Theory of Computing, Seattle (1974).

[Im 1] Imrich, W., Graphen mit transitiver Automorphismengruppe, Monatsh. Math., 73 (1969), 341-347.

[Im 2] Imrich, W., Watkins, M.E., On graphical regular representations of cyclic extensions of groups, Pacific J. Math., 55 (1974), 461-477.

[Iw 1] Iwahori, N., Generalized Tits systems (Bruhat decomposition) on $\mathfrak{p}$-adic semi-simple groups, Proc. Symp. Pure Math., AMS, vol. 9, Providence, 1966, 71-83.

[Jo 1] Jordan, C., Théorèmes sur les groupes primitifs, J. Math. (2), 16 (1871), 383-408.

[Ka 1] Kagno, J.N., Linear graphs of degree ≤ 6 and their groups, Amer. J. Math., 68 (1946), 505-520.

[Ka 2] Kaluzhnin, L.A., Klin, M.H., On some maximal subgroups of symmetric and alternating groups, (in Russian), Mat. Sbornik, 87 (1972), 91-121.

[Ka 3] Karp, R.M., Reducibility among combinatorial problems, Complexity of Computer Computations, (ed. by R.E. Miller and J.W. Thatcher), Plenum Press, New York, 1972, 85-103.

[Ke 1] Kelmans, A.K., Graphs with equal number of paths of length 2 between incident and non-incident pairs of vertices, (in Russian), Trudy Seminara po combinatornoy Mat., MGU, January 1967, in "Voprosy kibernetiky," Moscow, 1972.

[Kn 1] Knödel, W., Ein Verfahren zur Feststellung der Isomorphie von endlichen zusamenhangenden Graphen, Computing, 8 (1971), 329-334.

[Ku 1] Kuhn, H.W., On imprimitive substitution groups, Amer. J. Math., 26 (1904), 45-102.

[Le 1] Lehman, A.A., On automorphisms of some classes of graphs, (in Russian), Avtomatica y Telemecanica, 1970, no. 2, 75-82.

[Le 2] Levi, G., Graph isomorphism: A heuristic edge-partitioning-oriented algorithm, Computing, 12 (1974), 291-313.

[Li 1] Lichtenbaum, L.M., Eigenvalues of a simple graph, (in Russian), Trudy 3-vo Vsesoyuznovo Mat. Syezda, vol. 1, Leningrad, 1956, 135.

[Li 2] Lichtenbaum, L.M., A duality theorem for simple graphs, (in Russian), Uspehi Mat. Nauk, 13 (1958), no. 8, 185.

[Li 3] Lichtenbaum, L.M., Traces of powers of the adjacency matrix of a simple graph, Izvestiya VUZov, 1959, no. 5, 154-163.

[Mo 1] Moon, J.W., On the line graph of a complete bigraph, Ann. Math. Stat., 34 (1963), 664-667.

[Mo 2] Morgan, H.L., The generation of a unique machine description for chemical structures, J. Chem. Docum., 5 (1965), 107-112.

[Na 1] Nagle, J.F., On ordering and identifying undirected linear graphs, J. Math. Phys., 7 (1966), 1588-1592.

[Or 1] Ore, O., Theory of graphs, AMS Coll. Publ., vol. 38, Providence, 1962.

[Pa 1] Paulus, A.J.L., Conference matrices and graphs of order 26, Report Techn. Univ. Eindhoven, 73-WSK-06 (1973).

[Qu 1] Quirin, W., Extension of some results of Manning and Wielandt on primitive permutation groups, Math. Z., 123 (1971), 223-230.

[Re 1] Read, R.C., Corneil, D.G., The graph isomorphism disease, J. Graph Th., to appear.

[Ro 1] Rosenfeld, M.Z., Note on construction and properties of some classes of strongly regular graphs, (in Russian), Uspehi Mat. Nauk, 28 (1973), no. 3, 197-198.

[Ru 1] Rudvalis, A., (v,k,λ)-graphs and polarities of (v,k,λ)-designs, Math. Z., 120 (1971), 224-230.

[Sa 1] Sabidussi, G., On a class of fixed-point-free graphs, Proc. AMS, 9 (1958), 800-804.

[Sa 2] Sabidussi, G., Vertex-transitive graphs, Monatsch. Math., 68 (1964), 426-437.

[Sa 3] Saucier, G., Un algorithme efficace recherchant l'isomorphism de deux graphes, Rev. Frans. Inform. Rech. Oper., 5 (1971), R3, 39-51.

[Sc 1] Schur, I., Zur Theorie der einfach transitiven Permutations-gruppen, Sitzungber. Preuss. Akad. Wiss. (1933), 598-623.

[Se 1] Seidel, J.J., Strongly regular graphs of L_2-type and of triangular type, Indag. Math., 29 (1967), 188-196.

[Se 2] Seidel, J.J., Strongly regular graphs with $(-1,1,0)$-adjacency matrix having eigenvalue 3, Linear Algebra Appl., 1 (1968), 281-298.

[Se 3] Seidel, J.J., Strongly regular graphs, Recent Progress in Combinatorics, ed. by W.T. Tutte, N.Y.-London, 1969.

[Se 4] Seidel, J.J., A survey of two-graphs, Proc. Intern. Coll. Theorie Combinatorie, Acc. Naz. Lincei, Roma, 1973, to appear.

[Se 5] Seidel, J.J., Graphs and two-graphs, Proc. Fifth Southeastern Conf. on Combinatorics, Graph Theory and Computing, Florida Atlantic Univ., Boca Raton, 1974.

[Se 6] Serre, J.-P., Groupes algebriques et corps de classes, Paris, 1959.

[Sh 1] Shrikhande, S.S., On a characterization of the triangular association scheme, Ann. Math. Stat., 30 (1959), 39-47.

[Sh 2] Shrikhande, S.S., The uniqueness of the L_2-association scheme, Ann. Math. Stat., 30 (1959), 781-798.

[Sh 3] Shrikhande, S.S., Bhat, Vasanti N., Nonisomorphic solutions of pseudo-(3,5,2) and pseudo-(3,6,3) graphs, Ann. N.Y. Acad. Sci., 175 (1970), no. 1, 331-350.

[Sh 4] Shrikhande, S.S., Bhat, Vasanti N., Graphs derived from $L_3(5)$ graphs, Sankhyā, Series A, 33 (1971), 315-350.

[Sh 5] Shirkhande, S.S., Bhat, Vasanti N., Seidel-equivalence in $LB_3(6)$ graphs, Aequat. Math., 7 (1971), N 2/3.

[Si 1] Sims, C.C., Graphs and finite permutation groups I, Math. Z., 95 (1967), 76-86.

[Si 2] Sirovich, F., Isomorphismo fra graphi: un algorithmo efficiente per trovare tutti gli isomorphismi, Calcolo, 8 (1971), 301-337.

[Sk 1] Skorobogatov, V.A., On determination of isomorphisms of non-oriented graphs, (in Russian), "Vychislityelnyye Systemy," no. 33, Novosibirsk, 1969.

[St 1] Steen, J.-P., Principle d'un algorithme de recherche d'un isomorphisme entre deux graphes, Rev. Franc. Inform. Rech. Oper., 3 (1969), 51-69.

[Sw 1] Swift, J.D., Isomorph rejection in exhaustive search techniques, AMS Proc. Symp. Appl. Math., vol. 10, Providence, 1960, 195-200.

[Ta 1] Tamaschke, O., On permutation groups, Ann. Mat. Pura Appl., Ser. IV, 80 (1968), 235-279.

[Ta 2] Tamaschke, O., On the theory of Schur-rings, Ann. Mat. Pura Appl., Ser. IV, 81 (1969), 1-44.

[Ta 3] Tamaschke, O., On Schur-rings which define a proper character theory of finite groups, Math. Z., 117 (1970), 340-360.

[Ti 1] Tits, J., Groupes finis simplex sporadiques, Sem. Bourbaki, 1969 (1970), exp. 375, Lecture Notes in Math., no. 180, 1971, 187-211.

[Tu 1] Turner, J., Generalized matrix functions and the graph isomorphism problem, SIAM J. Appl. Math., 16 (1968), 520-526.

[Un 1] Unger, S.H., GIT-a heuristic program for testing pairs of directed line graphs for isomorphism, Comm. ACM, (1964), 26-34.

[Wa 1] Wallis, W.D., A non-existence theorem for (v,k,λ)-graphs, J. Austr. Math. Soc., 11 (1970), 381-383.

[Wa 2] Watkins, M.E., On the action of non-abelian groups on graphs, J. Comb. Theory, 11 (1971), 95-104.

[Wa 3] Watkins, M.E., Graphical representations of Alt_n, Sym_n et al., Aequat. Math., 11 (1974),

[We 1] Weil, A., Algebras with involutions and the classical groups, J. Indian Math. Soc., 24 (1960), 589-623.

[We 2] Weinberg, L., A simple and efficient algorithm for determining isomorphism of planar triply connected graphs, Trans. IEEE, CT 13 (1966), 142-148.

[We 3] Weisfeiler, B., Lehman, A.A., A reduction of a graph to a canonical form and an algebra arising during this reduction, (in Russian), Nauchno-Technicheskaya Informatsia, Seriya 2, 1968, no. 9, 12-16.

[Wh 1] Whitney, H., Congruent graphs and connectivity of graphs, Amer. J. Math., 54 (1932), 150-168.

[Wi 1] Wielandt, H., Permutation groups, Pasadena, 1965.

[Wi 2] Wielandt, H., Permutation groups through invariant relations and invariant functions, Ohio Univ., 1969.

[Wi 3] Wielandt, H., Permutation representation, Ill. J. Math., 13 (1969), 91-94.

[Yo 1] Yokonuma, T., Sur la structure des anneaux de Hecke d'un groupe de Chevalley fini, C.R. Acad. Sci., A 264 (1967), 344-347.

[Zy 1] Zykov, A.A., Theory of finite graphs, (in Russian), vol. 1, "Nauka," Novosibirsk, 1969.

DISTRIBUTION OF THE BIBLIOGRAPHY PER TOPICS

1. Isomorphism problem

Ba 1, Ba 2, Be 2, Co 3, Dj 1, Ha 1, Ho 9, Ho 10, Kn 1, Le 2, Li 3, Mo 2, Na 1, Sa 3, Si 2, Sk 1, St 1, Tu 1, Un 1, We 2, We 3, Wh 1, Re 1

2. Strongly regular graphs

Ad 2, Ah 1, Ai 1, Al 2, Ar 2, As 1, Be 1, Bo 2 - Bo 9, Br 1, Ca 2, Ch 1, Ch 2, Co 1, Co 2, Do 1, Do 2, Ge 1, Go 1, Go 2, He 2, Hi 1, Hi 4, Hi 5, Ho 1 - Ho 8, Ke 1, Mo 1, Pa 1, Ro 1, Ru 1, Se 1 - Se 5, Sh 1 - Sh 5, Ti 1, Wa 1

3. Cellular algebras and related formations

Bo 6, Ca 1, Fr 1, Fr 2, He 2, Hi 2, Hi 3, Hi 5, Sc 1, Si 1, Ta 1 - Ta 3, Wi 1 - Wi 3

4. Graphs with a given automorphism group

Ch 3, Ch 4, Da 1, Da 2, Fr 3, He 1, Im 1, Im 2, Ka 2, Le 1, Sa 1, Sa 2, Ti 1, Wa 2, Wa 3

5. Algorithmic questions (except isomorphism problem)

Ad 1, Ar 2, Br 2, Ka 3, Pa 1, Sw 1

INDEX OF NOTATIONS

(Con. i refers to subsection i in "Conventions, Assumptions, Notations")

General (all from Section "Conventions, Assumptions, Notations")

CELLULAR ALGEBRAS

SPECIAL CLASSES

NORMAL SUBCELLS AND FACTOR CELLS

ALGORITHMS

SUBJECT INDEX

(Con. i refers to subsection i of "Conventions, Assumptions, Notations")

Vol. 399: Functional Analysis and its Applications. Proceedings 1973. Edited by H. G. Garnir, K. R. Unni and J. H. Williamson. II, 584 pages. 1974.

Vol. 400: A Crash Course on Kleinian Groups. Proceedings 1974. Edited by L. Bers and I. Kra. VII, 130 pages. 1974.

Vol. 401: M. F. Atiyah, Elliptic Operators and Compact Groups. V, 93 pages. 1974.

Vol. 402: M. Waldschmidt, Nombres Transcendants. VIII, 277 pages. 1974.

Vol. 403: Combinatorial Mathematics. Proceedings 1972. Edited by D. A. Holton. VIII, 148 pages. 1974.

Vol. 404: Théorie du Potentiel et Analyse Harmonique. Edité par J. Faraut. V, 245 pages. 1974.

Vol. 405: K. J. Devlin and H. Johnsbråten, The Souslin Problem. VIII, 132 pages. 1974.

Vol. 406: Graphs and Combinatorics. Proceedings 1973. Edited by R. A. Bari and F. Harary. VIII, 355 pages. 1974.

Vol. 407: P. Berthelot, Cohomologie Cristalline des Schémas de Caracteristique p > o. II, 604 pages. 1974.

Vol. 408: J. Wermer, Potential Theory. VIII, 146 pages. 1974.

Vol. 409: Fonctions de Plusieurs Variables Complexes, Séminaire François Norguet 1970–1973. XIII, 612 pages. 1974.

Vol. 410: Séminaire Pierre Lelong (Analyse) Année 1972–1973. VI, 181 pages. 1974.

Vol. 411: Hypergraph Seminar. Ohio State University, 1972. Edited by C. Berge and D. Ray-Chaudhuri. IX, 287 pages. 1974.

Vol. 412: Classification of Algebraic Varieties and Compact Complex Manifolds. Proceedings 1974. Edited by H. Popp. V, 333 pages. 1974.

Vol. 413: M. Bruneau, Variation Totale d'une Fonction. XIV, 332 pages. 1974.

Vol. 414: T. Kambayashi, M. Miyanishi and M. Takeuchi, Unipotent Algebraic Groups. VI, 165 pages. 1974.

Vol. 415: Ordinary and Partial Differential Equations. Proceedings 1974. XVII, 447 pages. 1974.

Vol. 416: M. E. Taylor, Pseudo Differential Operators. IV, 155 pages. 1974.

Vol. 417: H. H. Keller, Differential Calculus in Locally Convex Spaces. XVI, 131 pages. 1974.

Vol. 418: Localization in Group Theory and Homotopy Theory and Related Topics. Battelle Seattle 1974 Seminar. Edited by P. J. Hilton. VI, 172 pages 1974.

Vol. 419: Topics in Analysis. Proceedings 1970. Edited by O. E. Lehto, I. S. Louhivaara, and R. H. Nevanlinna. XIII, 392 pages. 1974.

Vol. 420: Category Seminar. Proceedings 1972/73. Edited by G. M. Kelly. VI, 375 pages. 1974.

Vol. 421: V. Poénaru, Groupes Discrets. VI, 216 pages. 1974.

Vol. 422: J.-M. Lemaire, Algèbres Connexes et Homologie des Espaces de Lacets. XIV, 133 pages. 1974.

Vol. 423: S. S. Abhyankar and A. M. Sathaye, Geometric Theory of Algebraic Space Curves. XIV, 302 pages. 1974.

Vol. 424: L. Weiss and J. Wolfowitz, Maximum Probability Estimators and Related Topics. V, 106 pages. 1974.

Vol. 425: P. R. Chernoff and J. E. Marsden, Properties of Infinite Dimensional Hamiltonian Systems. IV, 160 pages. 1974.

Vol. 426: M. L. Silverstein, Symmetric Markov Processes. X, 287 pages. 1974.

Vol. 427: H. Omori, Infinite Dimensional Lie Transformation Groups. XII, 149 pages. 1974.

Vol. 428: Algebraic and Geometrical Methods in Topology, Proceedings 1973. Edited by L. F. McAuley. XI, 280 pages. 1974.

Vol. 429: L. Cohn, Analytic Theory of the Harish-Chandra C-Function. III, 154 pages. 1974.

Vol. 430: Constructive and Computational Methods for Differential and Integral Equations. Proceedings 1974. Edited by D. L. Colton and R. P. Gilbert. VII, 476 pages. 1974.

Vol. 431: Séminaire Bourbaki – vol. 1973/74. Exposés 436–452. IV, 347 pages. 1975.

Vol. 432: R. P. Pflug, Holomorphiegebiete, pseudokonvexe Gebiete und das Levi-Problem. VI, 210 Seiten. 1975.

Vol. 433: W. G. Faris, Self-Adjoint Operators. VII, 115 pages. 1975.

Vol. 434: P. Brenner, V. Thomée, and L. B. Wahlbin, Besov Spaces and Applications to Difference Methods for Initial Value Problems. II, 154 pages. 1975.

Vol. 435: C. F. Dunkl and D. E. Ramirez, Representations of Commutative Semitopological Semigroups. VI, 181 pages. 1975.

Vol. 436: L. Auslander and R. Tolimieri, Abelian Harmonic Analysis, Theta Functions and Function Algebras on a Nilmanifold. V, 99 pages. 1975.

Vol. 437: D. W. Masser, Elliptic Functions and Transcendence. XIV, 143 pages. 1975.

Vol. 438: Geometric Topology. Proceedings 1974. Edited by L. C. Glaser and T. B. Rushing. X, 459 pages. 1975.

Vol. 439: K. Ueno, Classification Theory of Algebraic Varieties and Compact Complex Spaces. XIX, 278 pages. 1975

Vol. 440: R. K. Getoor, Markov Processes: Ray Processes and Right Processes. V, 118 pages. 1975.

Vol. 441: N. Jacobson, PI-Algebras. An Introduction. V, 115 pages. 1975.

Vol. 442: C. H. Wilcox, Scattering Theory for the d'Alembert Equation in Exterior Domains. III, 184 pages. 1975.

Vol. 443: M. Lazard, Commutative Formal Groups. II, 236 pages. 1975.

Vol. 444: F. van Oystaeyen, Prime Spectra in Non-Commutative Algebra. V, 128 pages. 1975.

Vol. 445: Model Theory and Topoi. Edited by F. W. Lawvere, C. Maurer, and G. C. Wraith. III, 354 pages. 1975.

Vol. 446: Partial Differential Equations and Related Topics. Proceedings 1974. Edited by J. A. Goldstein. IV, 389 pages. 1975.

Vol. 447: S. Toledo, Tableau Systems for First Order Number Theory and Certain Higher Order Theories. III, 339 pages. 1975.

Vol. 448: Spectral Theory and Differential Equations. Proceedings 1974. Edited by W. N. Everitt. XII, 321 pages. 1975.

Vol. 449: Hyperfunctions and Theoretical Physics. Proceedings 1973. Edited by F. Pham. IV, 218 pages. 1975.

Vol. 450: Algebra and Logic. Proceedings 1974. Edited by J. N. Crossley. VIII, 307 pages. 1975.

Vol. 451: Probabilistic Methods in Differential Equations. Proceedings 1974. Edited by M. A. Pinsky. VII, 162 pages. 1975.

Vol. 452: Combinatorial Mathematics III. Proceedings 1974. Edited by Anne Penfold Street and W. D. Wallis. IX, 233 pages. 1975.

Vol. 453: Logic Colloquium. Symposium on Logic Held at Boston, 1972–73. Edited by R. Parikh. IV, 251 pages. 1975.

Vol. 454: J. Hirschfeld and W. H. Wheeler, Forcing, Arithmetic, Division Rings. VII, 266 pages. 1975.

Vol. 455: H. Kraft, Kommutative algebraische Gruppen und Ringe. III, 163 Seiten. 1975.

Vol. 456: R. M. Fossum, P. A. Griffith, and I. Reiten, Trivial Extensions of Abelian Categories. Homological Algebra of Trivial Extensions of Abelian Categories with Applications to Ring Theory. XI, 122 pages. 1975.

Vol. 457: Fractional Calculus and Its Applications. Proceedings 1974. Edited by B. Ross. VI, 381 pages. 1975.

Vol. 458: P. Walters, Ergodic Theory – Introductory Lectures. VI, 198 pages. 1975.

Vol. 459: Fourier Integral Operators and Partial Differential Equations. Proceedings 1974. Edited by J. Chazarain. VI, 372 pages. 1975.

Vol. 460: O. Loos, Jordan Pairs. XVI, 218 pages. 1975.

Vol. 461: Computational Mechanics. Proceedings 1974. Edited by J. T. Oden. VII, 328 pages. 1975.

Vol. 462: P. Gérardin, Construction de Séries Discrètes p-adiques. »Sur les séries discrètes non ramifiées des groupes réductifs déployés p-adiques«. III, 180 pages. 1975.

Vol. 463: H.-H. Kuo, Gaussian Measures in Banach Spaces. VI, 224 pages. 1975.

Vol. 464: C. Rockland, Hypoellipticity and Eigenvalue Asymptotics. III, 171 pages. 1975.

Vol. 465: Séminaire de Probabilités IX. Proceedings 1973/74. Edité par P. A. Meyer. IV, 589 pages. 1975.

Vol. 466: Non-Commutative Harmonic Analysis. Proceedings 1974. Edited by J. Carmona, J. Dixmier and M. Vergne. VI, 231 pages. 1975.

Vol. 467: M. R. Essén, The Cos $\pi\lambda$ Theorem. With a paper by Christer Borell. VII, 112 pages. 1975.

Vol. 468: Dynamical Systems – Warwick 1974. Proceedings 1973/74. Edited by A. Manning. X, 405 pages. 1975.

Vol. 469: E. Binz, Continuous Convergence on C(X). IX, 140 pages. 1975.

Vol. 470: R. Bowen, Equilibrium States and the Ergodic Theory of Anosov Diffeomorphisms. III, 108 pages. 1975.

Vol. 471: R. S. Hamilton, Harmonic Maps of Manifolds with Boundary. III, 168 pages. 1975.

Vol. 472: Probability-Winter School. Proceedings 1975. Edited by Z. Ciesielski, K. Urbanik, and W. A. Woyczyński. VI, 283 pages. 1975.

Vol. 473: D. Burghelea, R. Lashof, and M. Rothenberg, Groups of Automorphisms of Manifolds. (with an appendix by E. Pedersen) VII, 156 pages. 1975.

Vol. 474: Séminaire Pierre Lelong (Analyse) Année 1973/74. Edité par P. Lelong. VI, 182 pages. 1975.

Vol. 475: Répartition Modulo 1. Actes du Colloque de Marseille-Luminy, 4 au 7 Juin 1974. Edité par G. Rauzy. V, 258 pages. 1975. 1975.

Vol. 476: Modular Functions of One Variable IV. Proceedings 1972. Edited by B. J. Birch and W. Kuyk. V, 151 pages. 1975.

Vol. 477: Optimization and Optimal Control. Proceedings 1974. Edited by R. Bulirsch, W. Oettli, and J. Stoer. VII, 294 pages. 1975.

Vol. 478: G. Schober, Univalent Functions – Selected Topics. V, 200 pages. 1975.

Vol. 479: S. D. Fisher and J. W. Jerome, Minimum Norm Extremals in Function Spaces. With Applications to Classical and Modern Analysis. VIII, 209 pages. 1975.

Vol. 480: X. M. Fernique, J. P. Conze et J. Gani, Ecole d'Eté de Probabilités de Saint-Flour IV-1974. Edité par P.-L. Hennequin. XI, 293 pages. 1975.

Vol. 481: M. de Guzmán, Differentiation of Integrals in R^n. XII, 226 pages. 1975.

Vol. 482: Fonctions de Plusieurs Variables Complexes II. Séminaire François Norguet 1974–1975. IX, 367 pages. 1975.

Vol. 483: R. D. M. Accola, Riemann Surfaces, Theta Functions, and Abelian Automorphisms Groups. III, 105 pages. 1975.

Vol. 484: Differential Topology and Geometry. Proceedings 1974. Edited by G. P. Joubert, R. P. Moussu, and R. H. Roussarie. IX, 287 pages. 1975.

Vol. 485: J. Diestel, Geometry of Banach Spaces – Selected Topics. XI, 282 pages. 1975.

Vol. 486: S. Stratila and D. Voiculescu, Representations of AF-Algebras and of the Group U (∞). IX, 169 pages. 1975.

Vol. 487: H. M. Reimann und T. Rychener, Funktionen beschränkter mittlerer Oszillation. VI, 141 Seiten. 1975.

Vol. 488: Representations of Algebras, Ottawa 1974. Proceedings 1974. Edited by V. Dlab and P. Gabriel. XII, 378 pages. 1975.

Vol. 489: J. Bair and R. Fourneau, Etude Géométrique des Espaces Vectoriels. Une Introduction. VII, 185 pages. 1975.

Vol. 490: The Geometry of Metric and Linear Spaces. Proceedings 1974. Edited by L. M. Kelly. X, 244 pages. 1975.

Vol. 491: K. A. Broughan, Invariants for Real-Generated Uniform Topological and Algebraic Categories. X, 197 pages. 1975.

Vol. 492: Infinitary Logic: In Memoriam Carol Karp. Edited by D. W. Kueker. VI, 206 pages. 1975.

Vol. 493: F. W. Kamber and P. Tondeur, Foliated Bundles and Characteristic Classes. XIII, 208 pages. 1975.

Vol. 494: A Cornea and G. Licea. Order and Potential Resolvent Families of Kernels. IV, 154 pages. 1975.

Vol. 495: A. Kerber, Representations of Permutation Groups II. V, 175 pages. 1975.

Vol. 496: L. H. Hodgkin and V. P. Snaith, Topics in K-Theory. Two Independent Contributions. III, 294 pages. 1975.

Vol. 497: Analyse Harmonique sur les Groupes de Lie. Proceedings 1973–75. Edité par P. Eymard et al. VI, 710 pages. 1975.

Vol. 498: Model Theory and Algebra. A Memorial Tribute to Abraham Robinson. Edited by D. H. Saracino and V. B. Weispfenning. X, 463 pages. 1975.

Vol. 499: Logic Conference, Kiel 1974. Proceedings. Edited by G. H. Müller, A. Oberschelp, and K. Potthoff. V, 651 pages 1975.

Vol. 500: Proof Theory Symposion, Kiel 1974. Proceedings. Edited by J. Diller and G. H. Müller. VIII, 383 pages. 1975.

Vol. 501: Spline Functions, Karlsruhe 1975. Proceedings. Edited by K. Böhmer, G. Meinardus, and W. Schempp. VI, 421 pages. 1976.

Vol. 502: János Galambos, Representations of Real Numbers by Infinite Series. VI, 146 pages. 1976.

Vol. 503: Applications of Methods of Functional Analysis to Problems in Mechanics. Proceedings 1975. Edited by P. Germain and B. Nayroles. XIX, 531 pages. 1976.

Vol. 504: S. Lang and H. F. Trotter, Frobenius Distributions in GL_2-Extensions. III, 274 pages. 1976.

Vol. 505: Advances in Complex Function Theory. Proceedings 1973/74. Edited by W. E. Kirwan and L. Zalcman. VIII, 203 pages. 1976.

Vol. 506: Numerical Analysis, Dundee 1975. Proceedings. Edited by G. A. Watson. X, 201 pages. 1976.

Vol. 507: M. C. Reed, Abstract Non-Linear Wave Equations. VI, 128 pages. 1976.

Vol. 508: E. Seneta, Regularly Varying Functions. V, 112 pages. 1976.

Vol. 509: D. E. Blair, Contact Manifolds in Riemannian Geometry. VI, 146 pages. 1976.

Vol. 510: V. Poènaru, Singularités C^∞ en Présence de Symétrie. V, 174 pages. 1976.

Vol. 511: Séminaire de Probabilités X. Proceedings 1974/75. Edité par P. A. Meyer. VI, 593 pages. 1976.

Vol. 512: Spaces of Analytic Functions, Kristiansand, Norway 1975. Proceedings. Edited by O. B. Bekken, B. K. Øksendal, and A. Stray. VIII, 204 pages. 1976.

Vol. 513: R. B. Warfield, Jr. Nilpotent Groups. VIII, 115 pages. 1976.

Vol. 514: Séminaire Bourbaki vol. 1974/75. Exposés 453 – 470. IV, 276 pages. 1976.

Vol. 515: Bäcklund Transformations. Nashville, Tennessee 1974. Proceedings. Edited by R. M. Miura. VIII, 295 pages. 1976.